VOLUME ONE HUNDRED AND THIRTY EIGHT

Progress in Molecular Biology and Translational Science

Growth Hormone in Health and Disease

VOLUME ONE HUNDRED AND THIRTY EIGHT

Progress in MOLECULAR BIOLOGY AND TRANSLATIONAL SCIENCE

Growth Hormone in Health and Disease

Edited by

FELIPE F. CASANUEVA
CIBER Fisiopatología Obesidad y Nutrición, Instituto de Salud Carlos III, Spain

ELSEVIER

AMSTERDAM • BOSTON • HEIDELBERG • LONDON
NEW YORK • OXFORD • PARIS • SAN DIEGO
SAN FRANCISCO • SINGAPORE • SYDNEY • TOKYO
Academic Press is an imprint of Elsevier

Academic Press is an imprint of Elsevier
125 London Wall, London EC2Y 5AS, UK
525 B Street, Suite 1800, San Diego, CA 92101-4495, USA
50 Hampshire Street, 5th Floor, Cambridge, MA 02139, USA
The Boulevard, Langford Lane, Kidlington, Oxford OX5 1GB, UK

First edition 2016

ISBN: 978-0-12-804827-6
ISSN: 1877-1173

For information on all Academic Press publications
visit our website at http://store.elsevier.com/

CONTENTS

Section III: GH Deficiency

CONTRIBUTORS

Stefano Allasia
Division of Endocrinology, Diabetology and Metabolism, Department of Medical Sciences, University of Turin, Turin, Italy

Miriam Azaretzky
Department of Medicine, Endocrinology Unit, Hospital T. Alvarez, Buenos Aires, Argentina

Silvia Barja-Fernandez
Grupo Fisiopatología Endocrina; Pediatric Department, Universidad de Santiago de Compostela, Instituto de Investigacion Sanitaria de Santiago de Compostela (IDIS), Complexo Hospitalario Universitario de Santiago (CHUS/SERGAS), Spain; CIBER Fisiopatología Obesidad y Nutrición, Instituto de Salud Carlos III, Spain

Ignacio Bernabeu
Department of Endocrinology and Nutrition, Complejo Hospitalario Universitario de Santiago de Compostela, Servicio Gallego de Salud (SERGAS); Universidad de Santiago de Compostela, La Coruña, Spain

Hugo R. Boquete
Department of Medicine, Endocrinology Unit, Hospital T. Alvarez, Buenos Aires, Argentina

Michael Buchfelder
Department of Neurosurgery, University of Erlangen-Nürnberg, Erlangen, Germany

Felipe F. Casanueva
CIBER Fisiopatología Obesidad y Nutrición, Instituto de Salud Carlos III, Spain; Laboratorio de Endocrinología Molecular y Celular, Instituto de Investigación Sanitaria de Santiago de Compostela (IDIS), Complexo Hospitalario Universitario de Santiago (CHUS/SERGAS), Santiago de Compostela, Spain

Cecilia Castelao
Grupo Fisiopatología Endocrina, Instituto de Investigacion Sanitaria de Santiago de Compostela (IDIS), Complexo Hospitalario Universitario de Santiago (CHUS/SERGAS), Spain; CIBER Fisiopatología Obesidad y Nutrición, Instituto de Salud Carlos III, Spain

Ana B. Crujeiras
CIBER Fisiopatología Obesidad y Nutrición, Instituto de Salud Carlos III, Spain; Laboratorio de Endocrinología Molecular y Celular, Instituto de Investigación Sanitaria de Santiago de Compostela (IDIS), Complexo Hospitalario Universitario de Santiago (CHUS/SERGAS), Santiago de Compostela, Spain

Julian Feulner
Department of Neurosurgery, University of Erlangen-Nürnberg, Erlangen, Germany

Hugo L. Fideleff
Department of Medicine, Endocrinology Unit, Hospital T. Alvarez, Buenos Aires, Argentina

Cintia Folgueira
Grupo Fisiopatología Endocrina, Instituto de Investigacion Sanitaria de Santiago de Compostela (IDIS), Complexo Hospitalario Universitario de Santiago (CHUS/SERGAS), Spain; CIBER Fisiopatología Obesidad y Nutrición, Instituto de Salud Carlos III, Spain

Stefano Frara
Endocrinology, University of Brescia, Brescia, Italy

Ezio Ghigo
Division of Endocrinology, Diabetology and Metabolism, Department of Medical Sciences, University of Turin, Turin, Italy

Andrea Giustina
Endocrinology, University of Brescia, Brescia, Italy

Züleyha Karaca
Department of Endocrinology, Erciyes University Medical School, Kayseri, Turkey

Fahrettin Kelestimur
Department of Endocrinology, Erciyes University Medical School, Kayseri, Turkey

Anne Klibanski
Neuroendocrine Unit, Massachusetts General Hospital, Boston, Massachusetts, USA; Harvard Medical School, Boston, Massachusetts, USA

John J. Kopchick
Edison Biotechnology Institute, Ohio University, Athens, Ohio, USA; Department of Biomedical Sciences, Heritage College of Osteopathic Medicine, Ohio University, Athens, Ohio, USA

Fabio Lanfranco
Division of Endocrinology, Diabetology and Metabolism, Department of Medical Sciences, University of Turin, Turin, Italy

Rosaura Leis
Pediatric Department, Universidad de Santiago de Compostela, Instituto de Investigacion Sanitaria de Santiago de Compostela (IDIS), Complexo Hospitalario Universitario de Santiago (CHUS/SERGAS), Spain

Edward O. List
Edison Biotechnology Institute, Ohio University, Athens, Ohio, USA; Department of Specialty Medicine, Heritage College of Osteopathic Medicine, Ohio University, Athens, Ohio, USA

Filippo Maffezzoni
Endocrinology, University of Brescia, Brescia, Italy

Mónica Marazuela
Department of Endocrinology and Nutrition, Hospital Universitario la Princesa, Instituto de Investigación Princesa, Universidad Autónoma de Madrid, Madrid, Spain

Gherardo Mazziotti
Endocrinology, University of Brescia, Brescia, Italy

Giovanna Motta
Division of Endocrinology, Diabetology and Metabolism, Department of Medical Sciences, University of Turin, Turin, Italy

Ana M. Ramos-Leví
Department of Endocrinology and Nutrition, Hospital Universitario la Princesa, Instituto de Investigación Princesa, Universidad Autónoma de Madrid, Madrid, Spain

Miguel Sampedro-Núñez
Department of Endocrinology and Nutrition, Hospital Universitario la Princesa, Instituto de Investigación Princesa, Universidad Autónoma de Madrid, Madrid, Spain

Luisa M. Seoane
Grupo Fisiopatología Endocrina; Pediatric Department, Universidad de Santiago de Compostela, Instituto de Investigacion Sanitaria de Santiago de Compostela (IDIS), Complexo Hospitalario Universitario de Santiago (CHUS/SERGAS), Spain

Martha G. Suárez
Department of Medicine, Endocrinology Unit, Hospital T. Alvarez, Buenos Aires, Argentina

Fatih Tanrıverdi
Department of Endocrinology, Erciyes University Medical School, Kayseri, Turkey

Nicholas A. Tritos
Neuroendocrine Unit, Massachusetts General Hospital, Boston, Massachusetts, USA; Harvard Medical School, Boston, Massachusetts, USA

Kürşad Ünlühızarcı
Department of Endocrinology, Erciyes University Medical School, Kayseri, Turkey

Jonathan A. Young
Edison Biotechnology Institute, Ohio University, Athens, Ohio, USA; Department of Biological Sciences, Ohio University, Athens, Ohio, USA

FOREWORD

The Growth Hormone (GH) and its regulation and function are contemporary topics and the whole field has been reactivated with the arrival of long-acting GH molecules as well as new molecules to control GH excess to the clinical practice. Although considerable efforts and publications were devoted to this area in the past, the new formulations will raise new problems and re-open the discussion of past ones, which will make mandatory an update of the conceptual basis of the whole system. For these reasons, the present book is timely and highly needed. The contributions of the authors, all of them experts in the area and with substantial contributions to its insight, were divided into three main blocks, with three chapters each: the first one regarding regulation of GH secretion and action, the second concerning excessive GH secretion and the third one addressing the states of GH deficiency.

In the first group, the regulation of GH secretion has been thoroughly reviewed regarding the role of Ghrelin, as one of the potential main regulators of the GH discharge by the pituitary gland. First discovered as a GH secretagogue, ghrelin was rapidly identified as a key signal in the regulation of energy homeostasis. An aspect which caused surprise was the fact that ghrelin is a hormone that circulates in two different forms, the acylated and the unacylated one. Only the acylated form is active on GH regulation but both forms are implicated in metabolic activities, which reinforces the concept that GH is intimately connected with metabolism. As circulating ghrelin is produced mainly by the gastric tissue, it is not surprising that the second chapter appears devoted to the regulation of GH by the splanchnic area. This regulation occurs not only through hormonal production, but also through the unexpected contribution of the vagus nerve and the set of hormone receptors present in the splanchnic area. The role of these tissues that have been largely ignored in the past, appear, now, under new perspective. Finally, the action of the GH cannot be understood without the analysis of its receptor, which is widely distributed along a variety of tissues of the body and with diverse actions. Not only the understanding of its function was needed, but it was also very important that this insight conducted to the know how of disrupting the receptor function as a way for clinically control the GH excess.

The clinical problem of GH excess, which translates into the diseases of acromegaly and gigantism, was the centre of the second section of chapters. A critical review of the latest criteria of managing acromegaly and the recently published international guidelines, were the topic of the first chapter in this section. Despite the fact that considerable progress was accomplished in the last ten years, this chapter provides an update of the clinical criteria in use. New concepts on how mutations in the GH receptor could affect treatment are followed by a final chapter in the surgical approach to treat and control acromegaly.

The final section addresses the opposite problem, i.e., the states of GH deficiency and the clinical problems associated with such states. Essentially, GH deficiency in children that results in dwarfism or reduced growth of the patient, and the impact of GH deficiency on bone metabolism, are the targets of the two chapters of this section. Finally, the states of GH deficiency generated by severe concussion to the brain, or GH deficiency due to traumatic brain injury is addressed in the last chapter of this part of the book. The relevance of the GH and pituitary hormones associated with traumatic brain injury appear under a new and very relevant aspect in this chapter and the impact on contact sports or military personnel are, nowadays, under scrutiny, in addition to the burden of car accidents in the modern society.

In summary, these are a group of chapters that will provide to the reader an updated, concise and authoritative view of the basic mechanisms and regulation of the somatotroph axis.

Felipe F Casanueva, MD, PhD
Professor of Medicine

SECTION I

Basic Concepts

CHAPTER ONE

Ghrelin Actions on Somatotropic and Gonadotropic Function in Humans

Giovanna Motta, Stefano Allasia, Ezio Ghigo, Fabio Lanfranco[1]
Division of Endocrinology, Diabetology and Metabolism, Department of Medical Sciences, University of Turin, Turin, Italy
[1] Corresponding author: e-mail address: fabio.lanfranco@unito.it.

Contents

Abstract

Ghrelin, a 28 amino-acid octanoylated peptide predominantly produced by the stomach, was discovered to be the natural ligand of the type 1a GH secretagogue receptor (GHS-R1a). It was thus considered as a natural GHS additional to GHRH, although later on ghrelin has mostly been considered a major orexigenic factor. The GH-releasing action of ghrelin takes place both directly on pituitary cells and through modulation of GHRH from the hypothalamus; some functional antisomatostatin action has also been shown. However, ghrelin is much more than a natural GH secretagogue. In fact, it also modulates lactotroph and corticotroph secretion in humans as well as in animals and plays a relevant role in the modulation of the hypothalamic-pituitary-gonadal function. Several studies have indicated that ghrelin plays an inhibitory effect on gonadotropin pulsatility, is involved in the regulation of puberty onset in animals, and may regulate spermatogenesis, follicular development and ovarian cell functions in humans.

Progress in Molecular Biology and Translational Science, Volume 138
ISSN 1877-1173
http://dx.doi.org/10.1016/bs.pmbts.2015.11.001

In this chapter ghrelin actions on the GH/IGF-I and the gonadal axes will be revised. The potential therapeutic role of ghrelin as a treatment of catabolic conditions will also be discussed.

1. INTRODUCTION

Ghrelin, a 28 amino-acid octanoylated peptide, was first isolated from the rat stomach in 1999.[1] It is predominantly synthesized by the endocrine X/A-like cells in the gastric fundus, but also expressed by several other tissues such as bowel, pancreas, kidney, immune system, placenta, testis, lung, and hypothalamus.[1–3] The word ghrelin was derived from "ghre" and "relin" that mean, respectively, "to grow" in Proto-Indo-European languages and "release".[4,5] The human ghrelin gene is localized in chromosome 3, at locus 3p25-2, made up of four exons and three introns.

Kojima *et al.* identified ghrelin as the endogenous ligand for the type 1a growth hormone secretagogue receptor (GHS-R1a) and as a powerful stimulus for the release of growth hormone (GH).[1]

Circulating ghrelin exists in several forms: acylated form (AG) and unacylated form (UAG). The latter is the most abundant, does not bind GHS-R1a, and is devoid of any neuroendocrine action. Nevertheless UAG is an active peptide exerting metabolic as well as nonendocrine actions, including cardiovascular and antiproliferative effects.[6–8] Moreover, UAG has been demonstrated to play a beneficial role in pancreatic beta cell function and survival.[9] As UAG does not bind GHS-R1a, these actions are likely mediated by a GHS-R subtype.

The hydroxyl group at Ser 3 is esterified by *n*-octanoic acid by ghrelin O-acyltransferase (GOAT): this acylation is essential for hormone binding to the GHS-R1a, for the GH-releasing capacity and most likely for its other endocrine, orexigenic, and metabolic actions.[1,6,7] In fact, ghrelin and many synthetic GHS influence a number of biological actions: (1) exhibit hypothalamic activities that result in stimulation of PRL and ACTH secretion; (2) negatively influence the pituitary-gonadal axis both at the central and the peripheral level; (3) stimulate appetite and a positive energy balance; (4) influence sleep and behavior; (5) control gastric motility and acid secretion; (6) modulate cardiovascular function and immune function; (7) modulate pancreatic exocrine and endocrine functions and affect glucose and lipid homeostasis.[6,7,10]

The GH-releasing property was the first recognized effect of AG.[11] However ghrelin also modulates lactotroph and corticotroph secretion in humans as well as in animals.[6,7,12,13] AG significantly stimulates PRL secretion *invitro* and *invivo*. The magnitude of the PRL-releasing action of ghrelin in humans is far lower than that of dopaminergic antagonists and TRH but similar to that of arginine.[6,7,14] Moreover, the stimulatory effect of ghrelin and synthetic GHS on the hypothalamus-pituitary-adrenal (HPA) axis in humans is remarkable and similar to that of the administration of naloxone, vasopressin, and even CRH.[6,7,13,15] GHS do not stimulate ACTH release directly from pituitary cell cultures and their stimulatory effect on the HPA axis is lost after pituitary stalk section in pigs; thus, ghrelin stimulates the HPA axis via the CNS.[6,7] In fact, ghrelin is likely to act at the hypothalamic level via stimulation of either CRH or arginine-vasopressin (AVP).[16,17]

This review will specifically focus on the somatotropic and gonadotropic actions of acylated ghrelin.

2. GHRELIN ACTIONS ON SOMATOTROPIC AXIS

2.1 GH-Releasing Action

The GH-releasing property of ghrelin was its first recognized effect.[1,13] Ghrelin as well as synthetic GHS possess strong and dose-related GH -releasing activity, both *in vitro* and *in vivo*, more marked in humans than in animals.[1,6,7,10,18] On the other hand, UAG was found not to affect GH secretion.[13]

The GH-releasing effect of AG is mediated by actions on the pituitary and, mainly, within the hypothalamus, through a positive action on GHRH secreting neurons and a concomitant functional antagonism of somatostatin activity.[19] At the hypothalamic level, ghrelin and GHS act via mediation of GHRH-secreting neurons as indicated by evidence that passive immunization against GHRH, as well as pretreatment with GHRH antagonists, reduces their stimulatory effect on GH secretion.[20–22] Moreover, the GH-releasing effect of GHS is markedly reduced in animals with lesions of the pituitary stalk.[6]

Natural and synthetic GHS stimulate GH release from somatotroph cells *invitro*, probably by depolarizing the somatotroph membrane and by increasing the amount of GH secreted per cell.[23,24]

The GH response to ghrelin bolus has been shown to be more robust than the response after GHRH bolus[15,18,25] or hexarelin[15] and is synergistic with the GHRH response,[15,26–28] suggesting a potential therapeutic use of ghrelin as a GH secretagogue.[29] Moreover it was discovered that the somatotroph releasing effect of AG is refractory to the direct inhibitory effect of a short-term elevation of GH levels, while it is markedly inhibited in the presence of increased IGF-I levels induced by 4-day rhGH administration. This finding suggests the possibility of a selective IGF-I-mediated feedback.[19]

The GH-releasing effect of AG and GHS undergoes marked age-related variations, increasing at puberty, persisting similar in adulthood and decreasing with aging; variations in estrogenic levels, the reduced expression of the hypothalamic GHS receptors in the aged human brain, GHRH hypoactivity and somatostatinergic hyperactivity would explain these age-related changes.[7,30,31]

The GH releasing effect of AG is independent of gender, does not vary with the menstrual cycle,[32] and occurs in a dose-dependent manner.[26,28,29,33]

At variance with GHRH, the stimulatory effect of AG on GH secretion is reduced both in obese and in anorectic patients,[7,34,35] in polycystic ovary syndrome,[36] hyperthyroidism,[37,38] Cushing's disease,[39] and primary hyperparathyroidism.[40]

Moreover the GH response to ghrelin bolus is reduced by centrally acting cholinergic antagonism,[41] but is not affected by peripherally acting cholinergic blockade,[31] cholinergic agonist,[31,41] oxytocin,[42] dopamine receptor blockade or by the most important inhibitory inputs on GH secretion such as glucose, free fatty acids, and β-adrenergic agonists, all acting to increase the hypothalamic somatostatin release.[14,19]

In healthy postmenopausal women, estradiol or combination estradiol–progestin replacement increases GH secretion in response to a ghrelin bolus,[43,44] and estradiol replacement increases basal, but not pulsatile, GH secretion in response to a ghrelin infusion.[45]

AG, as well as synthetic GHS, could have diagnostic and therapeutic implications based on the strong and reproducible GH-releasing effects. Since a damage to the pituitary stalk or to the pituitary reduces the GH response to a ghrelin bolus,[46,47] ghrelin and GHS, particularly when combined with GHRH, could be used as a potent and reliable provocative test to evaluate the capacity of the pituitary to release GH for the diagnosis of GH deficiency.[6,48,49] Long-acting and orally active ghrelin analogs might represent an anabolic treatment in frail elderly subjects or in catabolic patients. At present, however, there is no definite evidence showing the therapeutic efficacy of ghrelin analogs as GH/IGF-I axis-mediated anabolic agents in humans.

2.2 Potential Uses of Ghrelin in GH Secretion Disorders

2.2.1 Obesity

Circulating GH levels are low in obesity and obese subjects have a blunted responsiveness to GH stimuli, which is reversible after weight loss.[50,51] Moreover, GH levels are negatively correlated with BMI and GH half-life, secretory amplitude, and pulsatility are reduced in obesity.[52,53]

GH has strong lipolytic effects[54] and administration of GH for 9 months in middle-aged men with abdominal/visceral obesity has been shown to decrease abdominal visceral fat[55] and total body fat. The administration of a ghrelin mimetic in obese adults has been suggested to be useful to potentiate the GH lipolytic effect. However, data available up to now are not encouraging: in fact, the administration of an oral ghrelin mimetic, MK-677, to healthy male obese adults for 2 months increased fat free mass but did not decrease total and visceral fat mass.[56] Moreover, a 1-year MK-677 treatment increased lean body mass because of a sustained activation of the GH axis, but did not change total fat mass or abdominal visceral fat in healthy nonobese older adults.[57] In fact, ghrelin shows an adipogenic effect[58] through activation of lipogenic pathways in the central nervous system: subcutaneous administration of ghrelin to rodents has been shown to increase body fat mass.[59] In conclusion, activating the GH axis via ghrelin administration in obese subjects is possible, but ghrelin has an adipogenic effect that makes it an unlikely candidate for the treatment of obesity in humans.[57]

On the other hand, ghrelin levels are reduced in obese subjects compared to normal bodyweight controls and an attenuated suppression of ghrelin after meals has been reported.[60] The latter evidence has been hypothesized to be responsible for the lack of satiety in obese subjects after small meals. If this hypothesis was correct, a suppression of appetite in obese subjects could be obtained antagonizing the ghrelin system.[61] Thus, several different approaches have been investigated in the attempt to target the ghrelin system to ameliorate obesity.[61,62] These include the antagonisation of GHS-R1a, the neutralization of ghrelin signal using the vaccine approach[63] or monoclonal antighrelin antibodies,[64] and the inhibition of GOAT enzyme.[65]

2.2.2 Cancer Cachexia

Cachexia has been defined as weight loss >5% over a 6-month period in the absence of simple starvation, a BMI <20 kg/m^2 and weight loss >2%, or as severe bodyweight, fat, and muscle loss, and increased protein catabolism due to underlying diseases.[66]

Anorexia/cachexia in cancer have been partly explained by elevated proinflammatory cytokines such as IFN-γ, TNF-a, IL-1b, and IL-6.[5,67]

In murine models of cancer cachexia ghrelin administration appears to successfully diminish cachexia, increase appetite, and preserve lean muscle mass.[5] These effects are attributed to the orexigenic neuropeptides Agouti-Related Peptide (AgRP) and Neuropeptide Y (NPY) and to anti-inflammatory effects of ghrelin, respectively.[68]

Several human studies have reported increased plasma ghrelin levels in individuals with low compared to those with normal or higher BMI.[69,70] A large Japanese study in a nonobese population of 638 subjects revealed an inverse relationship between ghrelin and age, BMI, waist circumference, fasting plasma glucose, and insulin levels among other variables.[71]

Ghrelin levels are elevated in many different human cancer types, with the exception of gastrointestinal malignancies,[72] probably because they affect the ghrelin–gastric secreting areas. Interestingly, ghrelin elevation in many of these patients is still associated with poor appetite and weight loss. This has led some authors to postulate a state of ghrelin resistance that cannot be overcome even by reactive increases in endogenous ghrelin production.[5,73] Nevertheless, the administration of supraphysiologic doses of exogenous ghrelin or ghrelin mimetics has been demonstrated to have a beneficial effect in this setting of ghrelin resistance. Few studies demonstrate that ghrelin or ghrelin mimetic administration in advanced incurable cancer and anorexia increases energy intake and appetite.[74,75] An increase in these patients' meal appreciation score after ghrelin treatment has also been described.[76]

Anamorelin- ONO-7643 (ANAM) is a novel, orally active, ghrelin receptor agonist in clinical development for the treatment of cancer cachexia.[77] It is found to be associated with significant appetite-enhancing activity and resultant improvements in bodyweight, lean body mass. However, further studies are needed to confirm the significant potential of ANAM in cancer anorexia–cachexia syndrome.[78]

In summary, the composite preclinical literature indicates beneficial effects of ghrelin-based intervention in cancer–cachexia models and with an increase in lean body mass. Preliminary clinical data show that ghrelin maintains its GH releasing and orexigenic effect in the setting of cancer. However, further investigations should evaluate the effects of ghrelin administration on tumor growth.[5]

2.2.3 AIDS Associated Cachexia

During the early periods of the human immune deficiency virus (HIV)/AIDS epidemic, cachexia was a common condition. Aberrations in GHRH-GH-IGF-I axis are common in the complex of HIV and AIDS, particularly in case of lipodystrophy which results in complications such as chronic inflammation, insulin resistance, lipid and metabolic abnormalities. The processes involved in lipodystrophy are related to the suppression of GH production. The mechanism of low GH levels is due to increased somatostatin tone and decreased ghrelin secretion. The GHRH analog Tesamorelin is the only therapeutic option, which is FDA approved, to reduce abdominal fat excess in patients with HIV-associated lipodystrophy.[79,80]

On the other hand, elevated GH and low IGF-I levels are present in AIDS wasting syndrome, suggesting GH resistance.[81] To date, no reports of ghrelin or GHS use in this clinical setting are available.[5]

2.2.4 Anorexia Nervosa

Anorexia nervosa (AN) is a severe psychiatric disorder affecting about 0.9% of women and 0.3% of men[82] and has the highest mortality rate of any mental disorder.[83]

Total ghrelin levels, mostly in the UAG form, and GH levels are higher than controls in AN[69,84,85] and refeeding leads to a decrease in the peptide levels.[86] Elevated ghrelin levels are probably due to a decreased postprandial decline or to a state of ghrelin resistance in these patients.[87] Higher ghrelin levels in AN are likely to represent an adaptive response in order to stimulate eating and thereby increase bodyweight and fat.[83]

Hotta *et al.* demonstrated that the intravenous administration of ghrelin twice a day for 14 days in four out of five patients with restrictive AN improves epigastric discomfort or constipation and increases the hunger score and daily energy intake compared with the pretreatment period. These results imply that ghrelin has the potential as a new treatment for AN.[5,88]

In contrast, Miljic et al. reported that single-dose continuous administration of ghrelin in 15 patients with AN for 5 h failed to affect appetite.[89] It is possible that a single infusion is not sufficient to counteract the many factors that play a role in AN (such as anxiety, depression, and obsessive–compulsive disorder), and that a longer lasting ghrelin administration is needed to induce appetite changes.[90]

Similar to other states of malnutrition, AN may lead to peripheral GH resistance and decreased IGF-I.[91,92] GH increases not only due to the effects of ghrelin but also due to the absence of negative inhibition by IGF-I on GH release.

Broglio *et al.* showed higher basal morning ghrelin and GH levels and lower IGF-I levels in AN compared to normal women. The GH response to GHRH in AN was significantly higher than in normal subjects, while the GH response to ghrelin was significantly lower.[35] This indicates that AN patients are not only GH resistant, but also ghrelin resistant.

There are no long-term studies on the treatment of restrictive AN with either ghrelin or ghrelin mimetics. Apart from the orexigenic effects, it is unclear if treatment with agents that activate the GHS-R1a would have any effect on AN by increasing GH levels further and possibly raising IGF-I.[5]

2.2.5 Ageing

Ghrelin decline with ageing has been demonstrated by several studies.[53,93] However, the pituitary ghrelin receptor content does not decline with age[94] and the secretory response of the pituitary to ghrelin and GH secretagogues in the elderly is maintained.[31]

Age-related sarcopenia refers to the loss of muscle mass and muscle strength that is associated with aging. A number of mechanisms have been reported in age-related sarcopenia, including decreased appetite, reduced levels of anabolic hormones such as GH and IGF-I, increased muscle cell apoptosis, and increased proinflammatory cytokines.[95]

Nass *et al.* investigated the effects of MK-677 versus placebo in 65 healthy and nonsarcopenic elderly subjects. Fat-free mass and appendicular skeletal muscle mass (lean limb) increased with MK-677 treatment, but there was no change in functional capacity or quality of life.[57] However, this study included mainly active healthy older adults and the results may not be applicable to the general elderly population.

The absence of functional improvement with GH therapy has also been described by other studies, indicating that increasing GH levels in elderly subjects is not sufficient to treat sarcopenia.[96] Future larger studies focusing on sarcopenic elderly individuals are warranted to determine if strength and functional capacity will respond to ghrelin treatment or if combination therapy (i.e., with nutritional supplements) will be effective.

3. GHRELIN ACTIONS ON THE GONADAL AXIS

3.1 General Effects

Ghrelin regulates the hypothalamus-pituitary-gonadal (HPG) axis acting both at the central and at the peripheral level.[69,97–99] Increasing evidence supports an inhibitory effect of ghrelin in the regulation of gonadotropin secretion.[100] On the opposite side, ghrelin has been shown to stimulate LH secretion from cultured pituitary cells from goldfish[101] and female carp.[102] All these effects are summarized in Table 1.

AG suppresses LH pulsatility in rodent, ovine, and primate models.[99,103–108] It has also been shown to decrease LH responsiveness to GnRH from the pituitary *in vitro*.[107] However, ghrelin infusion decreased LH pulse frequency but not pulse amplitude in adult ovariectomized rhesus monkeys, suggesting that ghrelin could inhibit the GnRH pulse activity.[106]

Ghrelin can suppress not only LH, but also FSH secretion in male and female rats and this effect may depend on the manner of ghrelin administration.[109,110]

Ghrelin regulation of gonadotropin secretion in humans has been investigated mainly in male subjects. While in the first published study[29] different dosages of ghrelin increased GH but did not affect LH concentrations in normal males, two studies in men showed a delay and a suppression in LH pulse amplitude following acute i.v. ghrelin administration[111] and an inhibitory effect of ghrelin infusion on LH pulsatility.[112] In particular, we showed

Table 1 Ghrelin Effects on GnRH, LH, and/or FSH in Different Models.

	Effect on		
Animal Studies	**Hypothalamus**	**Pituitary**	**References**
Ovariectomized rat	↓ GnRH	↓ LH	103
		Basal ↑ LH ↑ FSH GnRH-stimulated LH ↓	99, 107, 109
Sheep		↓ LH	108
Rhesus monkey		↓ LH	106
Goldfish		↑ LH	101
Carp		↑ LH	102
Human	↓ GnRH	↓ LH	111, 140
		GnRH-stimulated LH ↔	112

↔, not modified; ↓, decreased; ↑, increased.

that a prolonged AG infusion quantitatively and qualitatively inhibits LH but not FSH secretion in healthy young males.[112] Moreover, in contrast with *in vitro* data showing that ghrelin reduces the LH response to GnRH in rodents,[107] the LH response to GnRH in humans is not modified by the exposure to AG.[112] These findings are therefore against the hypothesis that ghrelin plays any direct inhibitory role on pituitary gonadotropic cells. As AG inhibits the LH response to naloxone in humans, this clearly points toward a CNS-mediated inhibitory action on the HPG axis.[112]

In addition, ghrelin decreases GnRH release by hypothalamic explants/fragments *ex vivo*,[107] reinforcing the contention of a complex mode of action of ghrelin with inhibitory effects at central level and direct stimulatory action on basal gonadotropin secretion. Whether ghrelin action on the GnRH pulse generator is conducted directly on GnRH neurons or through indirect regulatory pathways is yet to be determined.[113] Some evidences suggest that ghrelin indirectly decreases gonadotropin secretion acting on central NPY, AgRP, or orexin expression,[97,100,114] which exhibit inhibitory effect on LH secretion.[113] On the other side, Forbes *et al.* demonstrated that ghrelin administration significantly reduces LH pulsatility and suppresses kisspeptin mRNA expression in ovariectomized rats and suggested that down-regulation of kisspeptin expression may play a critical role in the transduction of ghrelin-induced suppression of the reproductive function often observed during caloric restriction.[115]

It is well known that ghrelin is an important signal of energy insufficiency. In fact AN, malnutrition, and cachexia are generally associated to hypogonadism that reflects a functional impairment of neuroendocrine mechanisms.[116] Metabolic factors have a major impact on ghrelin secretion regulation, and the pathophysiological conditions mentioned earlier are not by chance associated with ghrelin hypersecretion.[35,117] Thus, it seems reasonable to hypothesize that ghrelin hypersecretion could have a role in the functional hypogonadism in AN, malnutrition, and cachexia.

Ghrelin acts also on testicular steroidogenesis inhibiting both hCG- and cAMP-stimulated testosterone release by Leydig cells in a dose-dependent manner.[97] Ghrelin effects on plasma testosterone concentrations in rats depend on the nutritional status. Indeed, in fed rats, ghrelin administration induces a slight decrease in testis mass without detectable changes in circulating testosterone, whereas in food-restricted animals, where endogenous ghrelin levels are increased, exogenous ghrelin administration induces overt decrease in plasma testosterone.[118] Once again, elevated ghrelin levels could contribute to male reproductive axis alterations in situations of energy deficit.[119]

Ghrelin expression by Leydig cells in humans is inversely correlated with serum testosterone concentrations, but is not directly related to spermatogenesis. Thus, it has been suggested that steroidogenic dysfunction is associated with increased ghrelin expression in human testis.[114,120]

3.2 Effects on Male and Female Puberty

Ghrelin has been shown to be involved in the regulation of puberty onset.[121] In fact, ghrelin delays pubertal onset both in male and female rats, males appearing to be more sensitive than females.[107,122]

Repeated ghrelin injections in male rats during the pubertal transition significantly decreased serum LH and testosterone levels and partially delayed balano–preputial separation (an external signal of puberty).[107,111,123] This suggests that elevated ghrelin levels (a signal of energy insufficiency) not only inhibit LH secretion but might also delay the normal timing of puberty. This inhibitory effect of ghrelin on LH secretion is elicited not only by AG, but also by UAG, which is able to inhibit LH secretion in pubertal male rats via a GHSR-1a independent mechanism.[111]

The mechanisms whereby ghrelin exerts these modulatory actions on puberty onset remain to be fully characterized. It is possible that ghrelin inhibits hypothalamic GnRH secretion and pulse frequency, as demonstrated *in vitro* and *ex vivo*.[124,125] As previously mentioned, it is still unclear whether ghrelin exerts this effect directly on GnRH neurons or through indirect regulatory pathways.

A similar inhibitory action of ghrelin has been suggested in humans, who show a progressive decline in circulating ghrelin levels during puberty. This delaying effect would be caused by the inhibition of GnRH-secreting neurons[70,126] and the decrease in plasma ghrelin levels during puberty progression has been interpreted as a permissive signal of HPG axis maturation because of a favorable metabolic condition.

3.3 Ghrelin in Male Reproduction

Ghrelin is present in the human testis and particularly in Leydig and Sertoli cells but not in germ cells[127] and the expression of ghrelin in Leydig cells is related to the degree of cell differentiation.[113,128]

Ghrelin expression has been reported to be inversely related with serum testosterone levels in patients with normozoospermia, obstructive azoospermia, or varicocele suggesting that ghrelin may have an indirect effect on spermatogenesis.[120]

In contrast to human and rodent data, in adult sheep testis strong ghrelin immunostaining is evident not only in Leydig and Sertoli cells but also in germ cells, with an indication of increased ghrelin immunoreactivity in germ cells during the mitotic phases and the meiotic prophases of the spermatogenic cycle.[129]

GHS-R1a has been identified in human germ cells, mainly in pachytene spermatocytes, as well as in Leydig and Sertoli cells.[128]

Ghrelin could regulate spermatogenesis in an autocrine and/or a paracrine manner.[119] Ghrelin is able to inhibit the expression of the gene encoding testicular stem cell factor (SCF), a Sertoli cell product involved in Leydig cell development and survival,[130] both after intratesticular injection *in vivo* and *in vitro*.[131] Moreover, *in vivo* intratesticular ghrelin inhibits the proliferative rate of immature Leydig cells both during puberty development and after selective ablation of pre-existing mature Leydig cells by administration of ethylene dimethane sulfonate.[131]

3.4 Ghrelin in Female Reproduction

Ghrelin and/or GHSR-1a expression has been reported in the gonads of several mammalian[132] and nonmammalian species.[133] Ghrelin and ghrelin receptors (GHS-R1a and GHS-R1b) are present in the human ovaries, particularly in the hilus interstitial cells and in mature corpora lutea and follicular cells and throughout the ovarian surface epithelium.[134] GHS R1a and R1b have been demonstrated in human granulosa-luteal cells.[135] The presence of GHS-R1a in ovarian follicles and corpora lutea suggests a potential regulatory role of systemic and locally produced ghrelin in the direct control of follicular development and ovarian cell functions.

In murine models, Caminos *et al.* (2003) indicated for the first time that ghrelin mRNA levels significantly vary depending on the phase of the cycle, with the lowest expression levels in proestrus and maximum values in the diestrous (day 1) phase.[98] Such a cyclic profile of expression, with peak levels in the luteal stages, is suggestive of predominant expression of ghrelin in the corpora lutea of the current cycle.[98,121] Thus, ghrelin mRNA reaches its highest levels when the corpora lutea enters into the functional phase and remains lower during corpora luteal formation and regression.

Treatment with GnRH antagonists has been shown to be associated with a decrease in ovarian ghrelin mRNA levels, and further studies have shown that in the proestrous stage, ghrelin levels depend on formation of corpora lutea.[98,121,136]

In cultured human granulosa luteal cells ghrelin exerts an inhibitory dose-dependent effect on steroidogenesis (progesterone and estradiol production), in the absence or in the presence of hCG.[135] Such effect may explain the suppression of the reproductive axis function in case of food deprivation, where limited resources are allocated to major physiological processes.[137]

In addition, progesterone secretion by human luteal cells *in vitro* is inhibited by ghrelin, which decreases the release of luteotropic factors and stimulates the secretion of luteolytic factors, thereby participating in the negative control of human luteal function.[138]

Experimental studies by Messini *etal.* demonstrated for the first time the inability of a ghrelin bolus to affect basal and GnRH-induced LH and FSH secretion in women, suggesting that ghrelin does not play a major physiological role in gonadotropin secretion in female subjects.[139] However more recent studies of the same Authors have shown an inhibitory effect of submaximal doses of ghrelin on gonadotropin secretion in women, in particular in the late follicular phase of the cycle.[140]

3.5 Pregnancy

Both ghrelin and GHSR mRNAs have been detected in the morula and in more advanced stages of embryo development,[104,137] showing a role in embryo preimplantation and development.

Tanaka *et al.* (2003) have also documented strong ghrelin expression in human placenta during the first trimester, especially in extravillous trophoblasts on the tips of chorionic villi, whereas at term the hormone levels are undetectable.[69]

Additionally, GHSR-1 mRNA has been found in the decidua, and *invitro* studies have shown that ghrelin is able to enhance human endometrial stromal cell decidualization.[69] These findings support the hypothesis that ghrelin, together with other messengers (including cytokines, interleukins, sex steroids, and prostaglandins, which are released by the invading chorionic tissue), may be a chemical mediator (in a paracrine and autocrine manner) of the regulation of endometrial stromal cell differentiation, which is essential for embryo implantation and the maintenance of pregnancy.[137,141–143]

In addition, reduced ghrelin levels have been demonstrated in third trimester maternal plasma,[144,145] in response to the marked change in maternal energy intake, which further suggests that reduced ghrelin levels could not only reflect maternal energy intake, but also prepare the uterus for parturition, since ghrelin possesses a relaxant effect on the uterus.[146]

Ghrelin can cross the feto–placental barrier,[147] therefore a role in fetal development has been hypothesized. This has been confirmed by the finding that fetuses from mothers receiving chronic ghrelin treatment have significantly higher birthweight compared to newborns from saline-treated mothers, and that their growth is significantly favored even in conditions of restricted maternal food intake.[147]

This is consistent with the concept that maternal ghrelin affects fetal development by mechanisms which are relatively independent of increased maternal nutritional state.[147]

However, the ghrelin amounts in the fetus are not totally of maternal origin, since it can also be produced in the human fetus.[148] In fact, increased ghrelin levels have been found in fetuses with intrauterine growth restriction,[148] and recent studies have shown that ghrelin levels in cord blood of full-term neonates are negatively correlated with birthweight.[149] The hypothesis that high ghrelin levels in intrauterine growth restriction fetuses may represent a "hunger signal" is further supported by the finding of higher umbilical cord ghrelin plasma concentrations in small for gestational age (SGA) neonates, compared with appropriate for gestational age (AGA) and large for gestational age (LGA) neonates.[150]

These data confirm a potential important role for ghrelin in the fetal and neonatal energy balance, and in allowing fetal adaptation to an adverse intrauterine environment.[150]

REFERENCES

1. Kojima M, Hosoda H, Date Y, Nakazato M, Matsuo H, Kangawa K. Ghrelin is a growth-hormone acylated peptide from stomach. *Nature*. 1999;402:656–660.
2. Kojima M, Hosoda H, Matsuo H, Kangawa K. Ghrelin: discovery of the natural endogenous ligand for the growth hormone secretagogue receptor. *Trends Endocrinol Metab*. 2001;12:118–122.
3. Kojima M, Kangawa K. Ghrelin: structure and function. *Physiol Rev*. 2005;85:495–522.
4. Walker AK, Gong Z, Park WM, Zigman JM, Sakata I. Expression of Serum Retinol Binding Protein and Transthyretin within Mouse Gastric Ghrelin Cells. *PLoS One*. 2013;8(6):e 64882
5. Guillory B, Splenser A, Garcia J. The role of ghrelin in anorexia-cachexia syndromes. *Vitam Horm*. 2013;92:61–106.
6. Van der Lely AJ, Tschöp M, Heiman ML, Ghigo E. Biological, physiological, pathophysiological, and pharmacological aspects of ghrelin. *Endocr Rev*. 2004;25:426–457.

7. Ghigo E, Broglio F, Arvat E, Maccario M, Papotti M, Muccioli G. Ghrelin: more than a natural GH secretagogue and/or an orexigenic factor. *Clin Endocrinol.* 2005;62:1–17.
8. Togliatto G, Trombetta A, Dentelli P, Baragli A, Rosso A, Granata R, Ghigo D, Pegoraro L, Ghigo E, Brizzi MF. Unacylated ghrelin rescues endothelial progenitor cell function in individuals with type 2 diabetes. *Diabetes.* 2010;59:1016–1025.
9. Granata R, Baragli A, Settanni F, Scarlatti F, Ghigo E. Unraveling the role of the ghrelin gene peptides in the endocrine pancreas. *J Mol Endocrinol.* 2010;45(3):107–118.
10. Broglio F, Prodam F, Riganti F, Muccioli G, Ghigo E. Ghrelin: from somatotrope secretion to new perspectives in the regulation of peripheral metabolic functions. *Front Horm Res.* 2006;35:102–114.
11. Kojima M, Hosoda H, Date Y, Nakazato M, Matsuo H, Kangawa K. Ghrelin is a growth-hormone-releasing acylated peptide from stomach. *Nature.* 1999;402:656–660.
12. Kojima M, Kangawa K. Ghrelin: more than endogenous growth hormone secretagogue. *Ann NY Acad Sci.* 2010;1200:140–148.
13. Baragli A, Lanfranco F, Allasia S, Granata R, Ghigo E. Neuroendocrine and metabolic activities of ghrelin gene products. *Peptides.* 2011;32:2323–2332.
14. Messini CI, Dafopoulos K, Chalvatzas N, Georgoulias P, Anifandis G, Messinis IE. Effect of ghrelin and metoclopramide on prolactin secretion in normal women. *J Endocrinol Invest.* 2011;34:276–279.
15. Arvat E, Maccario M, Di Vito L, Broglio F, Benso A, Gottero C, Papotti M, Muccioli G, Dieguez C, Casanueva FF, Deghenghi R, Camanni F, Ghigo E. Endocrine activities of ghrelin, a natural growth hormone secretagogue (GHS), in humans: comparison and interactions with hexarelin, a nonnaturalpeptidyl GHS, and GH-releasing hormone. *J Clin Endocrinol Metab.* 2001;86:1169–1174.
16. Korbonits M, Little JA, Forsling ML, Tringali G, Costa A, Navarra P, Trainer PJ, Grossman AB. The effect of growth hormone secretagogues and neuropeptide Y on hypothalamic hormone release from acute rat hypothalamic explants. *J Neuroendocrinol.* 1999;11:521–528.
17. Mozid AM, Tringali G, Forsling ML, Hendricks MS, Ajodha S, Edwards R, Navarra P, Grossman AB, Korbonits M. Ghrelin is released from rat hypothalamic explants and stimulates corticotrophin-releasing hormone and arginine-vasopressin. *Horm Metab Res.* 2003;35:455–459.
18. Arvat E, Di Vito L, Broglio F, Papotti M, Muccioli G, Dieguez C, Casanueva FF, Deghenghi R, Camanni F, Ghigo E. Preliminary evidence that ghrelin, the natural GH secretagogue (GHS)-receptor ligand, strongly stimulates GH secretion in humans. *J Endocrinol Invest.* 2000;23:493–495.
19. Benso A, Gramaglia E, Olivetti I, Tomelini M, Gigliardi VR, Frara S, Calvi E, Belcastro S, St Pierre DH, Ghigo E, Broglio F. The GH-releasing effect of acylated ghrelin in normal subjects is refractory to GH acute auto-feedback but is inhibited after short-term GH administration inducing IGF1 increase. *Eur J Endocrinol.* 2013;168(4):509–514.
20. Clark RG, Carlsson MS, Trojnar J, Robinson IC. The effects of a growth hormone-releasing peptide and growth hormone releasing factor in conscious and anaesthetized rats. *J Neuroendocrinol.* 1989;1:249–255.
21. Bowers CY, Sartor AO, Reynolds GA, Badger TM. On the actions of the growth hormone-releasing hexapeptide, GHRP. *Endocrinology.* 1991;128:2027–2035.
22. Barkan AL, Dimaraki EV, Jessup SK, Symons KV, Ermolenko M, Jaffe CA. Ghrelin secretion in humans is sexually dimorphic, suppressed by somatostatin, and not affected by the ambient growth hormone levels. *J Clin Endocrinol Metab.* 2003;88:2180–2184.
23. Goth MI, Lyons CE, Canny BJ, Thorner MO. Pituitary adenylate cyclase activating polypeptide, growth hormone (GH)-releasing peptide and GH-releasing hormone stimulate GH release through distinct pituitary receptors. *Endocrinology.* 1992;130: 939–944.

24. Smith RG, Cheng K, Schoen WR, Pong SS, Hickey G, Jacks T, Butler B, Chan WW, Chaung LY, Judith F. A nonpeptidyl growth hormone secretagogue. *Science.* 1993;260: 1640–1643.
25. Broglio F, Benso A, Gottero C, Prodam F, Grottoli S, Tassone F, Maccario M, Casanueva FF, Dieguez C, Deghenghi R, Ghigo E, Arvat E. Effects of glucose, free fatty acids or arginine load on the GH-releasing activity of ghrelin in humans. *Clin Endocrinol.* 2002;57:265–271.
26. Hataya Y, Akamizu T, Takaya K, Kanamoto N, Ariyasu H, Saijo M, Moriyama K, Shimatsu A, Kojima M, Kangawa K, Nakao K. A low dose of ghrelin stimulates growth hormone (GH) release synergistically with GH-releasing hormone in humans. *J Clin Endocrinol Metab.* 2001;86:4552.
27. Veldhuis JD, Norman C, Miles JM, Bowers CY. Sex steroids, GHRH, somatostatin, IGF-I, and IGFBP-1 modulate ghrelin's dose-dependent drive of pulsatile GH secretion in healthy older men. *J Clin Endocrinol Metab.* 2012;97:4753–4760.
28. Garin MC, Burns CM, Kaul S, Cappola AR. The human experience with ghrelin administration. *J Clin Endocrinol Metab.* 2013;98:1826–1837.
29. Takaya K, Ariyasu H, Kanamoto N, Iwakura H, Yoshimoto A, Harada M, Mori K, Komatsu Y, Usui T, Shimatsu A, Ogawa Y, Hosoda K, Akamizu T, Kojima M, Kangawa K, Nakao K. Ghrelin strongly stimulates growth hormone release in humans. *J Clin Endocrinol Metab.* 2000;85:4908–4911.
30. Ghigo E, Arvat E, Giordano R, Broglio F, Gianotti L, Maccario M, Bisi G, Graziani A, Papotti M, Muccioli G, Deghenghi R, Camanni F. Biologic activities of growth hormone secretagogues in humans. *Endocrine.* 2001;14:87–93.
31. Broglio F, Benso A, Castiglioni C, Gottero C, Prodam F, Destefanis S, Gauna C, Van Der Lely AJ, Deghenghi R, Bo M, Arvat E, Ghigo E. The endocrine response to ghrelin as a function of gender in humans in young and elderly subjects. *J Clin Endocrinol Metab.* 2003;88:1537–1542.
32. Messini CI, Dafopoulos K, Malandri M, Georgoulias P, Anifandis G, Messinis IE. Growth hormone response to submaximal doses of ghrelin remains unchanged during the follicular phase of the cycle. *Reprod Biol Endocrinol.* 2013;10(11):36.
33. Peino R, Baldelli R, Rodriguez-Garcia J, Rodriguez-Segade S, Kojima M, Kangawa K, Arvat E, Ghigo E, Dieguez C, Casanueva FF. Ghrelin-induced growth hormone secretion in humans. *Eur J Endocrinol.* 2000;143:R11–R14.
34. Tassone F, Broglio F, Destefanis S, Rovere S, Benso A, Gottero C, Prodam F, Rossetto R, Gauna C, Van der Lely AJ, Ghigo E, Maccario M. Neuroendocrine and metabolic effects of acute ghrelin administration in human obesity. *J Clin Endocrinol Metab.* 2003;88:5478–5483.
35. Broglio F, Gianotti L, Destefanis S, Fassino S, AbbateDaga G, Mondelli V, Lanfranco F, Gottero C, Gauna C, Hofland L, Van der Lely AJ, Ghigo E. The endocrine response to acute ghrelin administration is blunted in patients with anorexia nervosa, a ghrelin hypersecretory state. *Clin Endocrinol.* 2004;60:592–599.
36. Fusco A, Bianchi A, Mancini A, Milardi D, Giampietro A, Cimino V, Porcelli T, Romualdi D, Guido M, Lanzone A, Pontecorvi A, De Marinis L. Effects of ghrelin administration on endocrine and metabolic parameters in obese women with polycystic ovary syndrome. *J Endocrinol Invest.* 2007;30:948–956.
37. Nascif SO, Correa-Silva SR, Silva MR, Lengyel AM. Decreased ghrelin-induced GH release in thyrotoxicosis: comparison with GH releasing peptide-6 (GHRP-6) and GHRH. *Pituitary.* 2007;10:27–33.
38. Molica P, Nascif SO, Correa-Silva SR, de Sa LB, Vieira JG, Lengyel AM. Effects of ghrelin, GH-releasing peptide-6 (GHRP-6) and GHRH on GH, ACTH and cortisol release in hyperthyroidism before and after treatment. *Pituitary.* 2010;13: 315–323.

39. Correa-Silva SR, Nascif SO, Lengyel AM. Decreased GH secretion and enhanced ACTH and cortisol release after ghrelin administration in Cushing's disease: comparison with GH-releasing peptide-6 (GHRP-6) and GHRH. *Pituitary.* 2006;9:101–107.
40. Cecconi E, Bogazzi F, Morselli LL, Gasperi M, Procopio M, Gramaglia E, Broglio F, Giovannetti C, Ghigo E, Martino E. Primary hyperparathyroidism is associated with marked impairment of GH response to acylated ghrelin. *Clin Endocrinol.* 2008;69:197–201.
41. Maier C, Schaller G, Buranyi B, Nowotny P, Geyer G, Wolzt M, Luger A. The cholinergic system controls ghrelin release and ghrelin-induced growth hormone release in humans. *J Clin Endocrinol Metab.* 2004;89:4729–4733.
42. Coiro V, Volpi R, Stella A, Cataldo S, Chiodera P. Oxytocin does not modify GH, ACTH, cortisol and prolactin responses to ghrelin in normal men. *Neuropeptides.* 2011;45:139–142.
43. Veldhuis JD, Keenan DM, Iranmanesh A, Mielke K, Miles JM, Bowers CY. Estradiol potentiates ghrelin-stimulated pulsatile growth hormone secretion in postmenopausal women. *J Clin Endocrinol Metab.* 2006;91:3559–3565.
44. Villa P, Costantini B, Perri C, Suriano R, Ricciardi L, Lanzone A. Estro-progestin supplementation enhances the growth hormone secretory responsiveness to ghrelin infusion in postmenopausal women. *Fertil Steril.* 2008;89:398–403.
45. Kok P, Paulo RC, Cosma M, Mielke KL, Miles JM, Bowers CY, Veldhuis JD. Estrogen supplementation selectively enhances hypothalamo-pituitary sensitivity to ghrelin in postmenopausal women. *J Clin Endocrinol Metab.* 2008;93:4020–4026.
46. Maghnie M, Pennati MC, Civardi E, Di Iorgi N, Aimaretti G, Foschini ML, Corneli G, Tinelli C, Ghigo E, Lorini R, Loche S. GH response to ghrelin in subjects with congenital GH deficiency: evidence that ghrelin action requires hypothalamic-pituitary connections. *Eur J Endocrinol.* 2007;156:449–454.
47. Popovic V, Miljic D, Micic D, Damjanovic S, Arvat E, Ghigo E, Dieguez C, Casanueva FF. Ghrelin main action on the regulation of growth hormone release is exerted at hypothalamic level. *J Clin Endocrinol Metab.* 2003;88:3450–3453.
48. Ghigo E, Arvat E, Aimaretti G, Broglio F, Giordano R, Camanni F. Diagnostic and therapeutic uses of growth hormone-releasing substances in adult and elderly subjects. *Baillieres Clin Endocrinol Metab.* 1998;12:341–358.
49. Gasco V, Beccuti G, Baldini C, Prencipe N, Di Giacomo S, Berton A, Guaraldi F, Tabaro I, Maccario M, Ghigo E, Grottoli S. Acylated ghrelin as a provocative test for the diagnosis of GH deficiency in adults. *Eur J Endocrinol.* 2012;168:23–30.
50. Williams T, Berelowitz M, Joffe SN, Thorner MO, Rivier J, Vale W, Frohman LA. Impaired growth hormone releasing factor in obesity. *N Engl J Med.* 1984;311:1403–1407.
51. Maccario M, Grottoli S, Procopio M, Oleandri SE, Rossetto R, Gauna C, Arvat E, Ghigo E. The GH/IGF-I axis in obesity: influence of neuro-endocrine and metabolic factors. *Int J Obes Relat Metab Disord.* 2000;24(Suppl 2):S96–S99.
52. Scacchi M, Pincelli AI, Cavagnini F. Growth hormone in obesity. *Int J Obes Relat Metab Disord.* 1999;23:260–271.
53. Nass R, Gaylinn BD, Thorner MO. The role of ghrelin in GH secretion and GH disorders. *Mol Cell Endocrinol.* 2011;340:10–14.
54. Nass R, Thorner MO. Impact of the GH-cortisol ratio on the age-dependent changes in body composition. *Growth Horm IGF Res.* 2002;12:147–161.
55. Johannsson G, Mårin P, Lönn L, Ottosson M, Stenlöf K, Björntorp P, Sjöström L, Bengtsson BA. Growth hormone treatment of abdominally obese men reduces abdominal fat mass, improves glucose and lipoprotein metabolism, and reduces diastolic blood pressure. *J Clin Endocrinol Metab.* 1997;82:727–734.

56. Svensson J, Lönn L, Jansson JO, Murphy G, Wyss D, Krupa D, Cerchio K, Polvino W, Gertz B, Boseaus I, Sjöström L, Bengtsson BA. Two-month treatment of obese subjects with the oral growth hormone (GH) secretagogue MK-677 increases GH secretion, fat-free mass, and energy expenditure. *J Clin Endocrinol Metab.* 1998;83: 362–369.
57. Nass R, Pezzoli SS, Oliveri MC, Patrie JT, Harrell Jr FE, Clasey JL, Heymsfield SB, Bach MA, Vance ML, Thorner MO. Effects of an oral ghrelin mimetic on body composition and clinical outcomes in healthy older adults: a randomized trial. *Ann Intern Med.* 2008;149:601–611.
58. Cowley MA, Smith RG, Diano S, Tschöp M, Pronchuk N, Grove KL, Strasburger CJ, Bidlingmaier M, Esterman M, Heiman ML, Garcia-Segura LM, Nillni EA, Mendez P, Low MJ, Sotonyi P, Friedman JM, Liu H, Pinto S, Colmers WF, Cone RD, Horvath TL. The distribution and mechanism of action of ghrelin in the CNS demonstrates a novel hypothalamic circuit regulating energy homeostasis. *Neuron.* 2003;37:649–661.
59. Tschöp M, Smiley DL, Heiman ML. Ghrelin induces adiposity in rodents. *Nature.* 2000;407:908–913.
60. Le Roux CW, Patterson M, Vincent RP, Hunt C, Ghatei MA, Bloom SR. Postprandial plasma ghrelin is suppressed proportional to meal calorie content in normal weight but not obese subjects. *J Clin Endocrinol Metab.* 2005;90:1068–1071.
61. Patterson M, Bloom SR, Gardiner JV. Ghrelin and appetite control in humans-potential application in the treatment of obesity. *Peptides.* 2011;32:2290–2294.
62. Sato T, Ida T, Nakamura Y, Shiimura Y, Kangawa K, Kojima M. Physiological roles of ghrelin on obesity. *Obes Res Clin Pract.* 2014;8(5):e405–e413.
63. Zorrilla EP, Iwasaki S, Moss JA, Chang J, Otsuji J, Inoue K, Meijler MM, Janda KD. Vaccination against weight gain. *Proc Natl Acad Sci USA.* 2006;103:13226–13231.
64. Lu SC, Xu J, Chinookoswong N, Liu S, Steavenson S, Gegg C, Brankow D, Lindberg R, Véniant M, Gu W. An acyl-ghrelin specific neutralizing antibody inhibits the acute ghrelin-mediated orexigenic effects in mice. *Mol Pharmacol.* 2009;75:901–907.
65. Barnett BP, Hwang Y, Taylor MS, Kirchner H, Pfluger PT, Bernard V, Lin YY, Bowers EM, Mukherjee C, Song WJ, Longo PA, Leahy DJ, Hussain MA, Tschöp MH, Boeke JD, Cole PA. Glucose and weight control in mice with a designed ghrelin O-acyltransferase inhibitor. *Science.* 2010;330:1689–1692.
66. Muscaritoli M, Anker SD, Argilés J, Aversa Z, Bauer JM, Biolo G, Boirie Y, Bosaeus I, Cederholm T, Costelli P, Fearon KC, Laviano A, Maggio M, Rossi Fanelli F, Schneider SM, Schols A, Sieber CC. Consensus definition of sarcopenia, cachexia and pre-cachexia: joint document elaborated by Special Interest Groups (SIG) "cachexia-anorexia in chronic wasting diseases" and "nutrition in geriatrics". *Clin Nutr.* 2010;29:154–159.
67. Guney Y, OzelTurkcu U, Hicsonmez A, NalcaAndrieu M, Kurtman C. Ghrelin may reduce radiation-induced mucositis and anorexia in head-neck cancer. *Med Hypotheses.* 2007;68:538–540.
68. Wang L, Basa NR, Shaikh A, Luckey A, Heber D, St-Pierre DH, Taché Y. LPS inhibits fasted plasma ghrelin levels in rats: role of IL-1 and PGs and functional implications. *Am J Physiol Gastrointest Liver Physiol.* 2006;29:G611–G620.
69. Tanaka M, Naruo T, Nagai N, Kuroki N, Shiiya T, Nakazato M, Matsukura S, Nozoe S. Habitual binge/purge behavior influences circulating ghrelin levels in eating disorders. *J Psychiatr Res.* 2003;37:17–22.
70. Soriano-Guillén L, Barrios V, Campos-Barros A, Argente J. Ghrelin levels in obesity and anorexia nervosa: effect of weight reduction or recuperation. *J Pediatr.* 2004;144: 36–42.

71. Nanjo Y, Adachi H, Hirai Y, Enomoto M, Fukami A, Otsuka M, Yoshikawa K, Yokoi K, Ogata K, Tsukagawa E, Kasahara A, Murayama K, Yasukawa H, Kojima M, Imaizumi T. Factors associated with plasma ghrelin level in Japanese general population. *Clin Endocrinol.* 2011;74:453–458.
72. Kemik O, Sumer A, Kemik AS, Hasirci I, Purisa S, Dulger AC, Demiriz B, Tuzun S. The relationship among acute-phase response proteins, cytokines and hormones in cachectic patients with colon cancer. *World J Surg Oncol.* 2010;8:85.
73. Garcia JM, Garcia-Touza M, Hijazi RA, Taffet G, Epner D, Mann D, Smith RG, Cunningham GR, Marcelli M. Active ghrelin levels and active to total ghrelin ratio in cancer-induced cachexia. *J Clin Endocrinol Metab.* 2005;90:2920–2926.
74. Strasser F, Lutz TA, Maeder MT, Thuerlimann B, Bueche D, Tschöp M, Kaufmann K, Holst B, Brändle M, von Moos R, Demmer R, Cerny T. Safety, tolerability and pharmacokinetics of intravenous ghrelin for cancer-related anorexia/cachexia: a randomised, placebo-controlled, double-blind, double-crossover study. *Br J Cancer.* 2008;98:300–308.
75. Lundholm K, Gunnebo L, Körner U, Iresjö BM, Engström C, Hyltander A, Smedh U, Bosaeus I. Effects by daily long term provision of ghrelin to unselected weight-losing cancer patients: a randomized double-blind study. *Cancer.* 2011;116: 2044–2052.
76. Dickson SL, Egecioglu E, Landgren S, Skibicka KP, Engel JA, Jerlhag E. The role of the central ghrelin system in reward from food and chemical drugs. *Mol Cell Endocrinol.* 2011;340:80–87.
77. Garcia JM, Polvino WJ. Effect on body weight and safety of RC-1291, a novel, orally available ghrelin mimetic and growth hormone secretagogue: results of a phase I, randomized, placebo-controlled, multiple-dose study in healthy volunteers. *Oncologist.* 2007;12:594–600.
78. Pietra C, Takeda Y, Tazawa-Ogata N, Minami M, Yuanfeng X, Duus EM, Northrup R. Anamorelin HCl (ONO-7643), a novel ghrelin receptor agonist, for the treatment of cancer anorexia-cachexia syndrome: preclinical profile. *J Cachexia Sarcopenia Muscle.* 2014;5:329–337.
79. Stanley TL, Feldpausch MN, Oh J, Branch KL, Lee H, Torriani M, Grinspoon SK. Effect of tesamorelin on visceral fat and liver fat in HIV-infected patients with abdominal fat accumulation: a randomized clinical trial. *JAMA.* 2014;312:380–389.
80. Gullett NP, Hebbar G, Ziegler TR. Update on clinical trials of growth factors and anabolic steroids in cachexia and wasting. *Am J Clin Nutr.* 2010;91:1143S–1147S.
81. Jain S, Desai N, Bhangoo A. Pathophysiology of GHRH-growth hormone-IGF1 axis in HIV/AIDS. *Rev Endocr Metab Disord.* 2013;14:113–118.
82. Hudson JI, Hiripi E, Pope Jr HG, Kessler RC. The prevalence and correlates of eating disorders in the National Comorbidity Survey Replication. *Biol Psychiatry.* 2007;61:348–358.
83. Atalayer D, Gibson C, Konopacka A, Geliebter A. Ghrelin and eating disorders. *Prog Neuropsychopharmacol Biol Psychiatry.* 2013;40:70–82.
84. Misra M, Miller KK, Kuo K, Griffin K, Stewart V, Hunter E, Herzog DB, Klibanski A. Secretory dynamics of leptin in adolescent girls with anorexia nervosa and healthy adolescents. *Am J Physiol Endocrinol Metab.* 2005;289:E373–E381.
85. Uehara M, Yasuhara D, Nakahara T, Harada T, Koyama KI, Ushikai M, Asakawa A, Inui A. Increase in energy intake leads to a decrease in obestatin in restricting-type ofanorexia nervosa. *Exp Clin Endocrinol Diabetes.* 2011;119:536–539.
86. Koyama KI, Yasuhara D, Nakahara T, Harada T, Uehara M, Ushikai M, Asakawa A, Inui A. Changes in acyl ghrelin, des-acyl ghrelin, and ratio of acyl ghrelin to total ghrelin with short-term refeeding in female inpatients with restricting-type anorexia nervosa. *Horm Metab Res.* 2010;42:595–598.

87. Otto B, Cuntz U, Fruehauf E, Wawarta R, Folwaczny C, Riepl RL, Heiman ML, Lehnert P, Fichter M, Tschöp M. Weight gain decreases elevated plasma ghrelin concentrations of patients with anorexia nervosa. *EurJ Endocrinol.* 2001;145:669–673.
88. Hotta M, Ohwada R, Akamizu T, Shibasaki T, Takano K, Kangawa K. Ghrelin increases hunger and food intake in patients with restricting-type anorexia nervosa: a pilot study. *EndocrJ.* 2009;56:1119–1128.
89. Miljic D, Pekic S, Djurovic M, Doknic M, Milic N, Casanueva FF, Ghatei M, Popovic V. Ghrelin has partial or no effect on appetite, growth hormone, prolactin, and cortisol release in patients with anorexia nervosa. *J Clin Endocrinol Metab.* 2006;91:1491–1495.
90. Ogiso K, Asakawa A, Amitani H, Inui A. Ghrelin and anorexia nervosa: a psychosomatic perspective. *Nutrition.* 2011;27:988–993.
91. Gianotti L, Lanfranco F, Ramunni J, Destefanis S, Ghigo E, Arvat E. GH/IGF-I axis in anorexia nervosa. *EatWeight Disord.* 2002;7:94–105.
92. Terashi M, Asakawa A, Harada T, Ushikai M, Coquerel Q, Sinno MH, Déchelotte P, Inui A, Fetissov SO. Ghrelin reactive autoantibodies in restrictive anorexia nervosa. *Nutrition.* 2011;27:407–413.
93. Rigamonti AE, Pincelli AI, Corrà B, Viarengo R, Bonomo SM, Galimberti D, Scacchi M, Scarpini E, Cavagnini F, Müller EE. Plasma ghrelin concentrations in elderly subjects: comparison with anorexic and obese patients. *J Endocrinol.* 2002;175:R1–R5.
94. Sun Y, Garcia JM, Smith RG. Ghrelin and growth hormone secretagogue receptor expression in mice during aging. *Endocrinology.* 2007;148:1323–1329.
95. Hall DT, Ma JF, Marco SD, Gallouzi IE. Inducible nitric oxide synthase(iNOS) in muscle wasting syndrome, sarcopenia, and cachexia. *Aging.* 2011;3:702–715.
96. Nass R, Johannsson G, Christiansen JS, Kopchick JJ, Thorner MO. The aging population-is there a role for endocrine interventions? *Growth Horm IGF Res.* 2009;19:89–100.
97. Tena-Sempere M, Barreiro ML, Gonzalez LC, Gaytan F, Zhang FP, Caminos JE, Pinilla L, Casanueva FF, Dieguez C, Aguilar E. Novel expression and functional role of ghrelin in rat testis. *Endocrinology.* 2002;143:717–725.
98. Caminos JE, Tena-Sempere M, Gaytán F, Sanchez-Criado JE, Barreiro ML, Nogueiras R, Casanueva FF, Aguilar E, Diéguez C. Expression of ghrelin in the cyclic and pregnant rat ovary. *Endocrinology.* 2003;144:1594–1602.
99. Fernández-Fernández R, Tena-Sempere M, Aguilar E, Pinilla L. Ghrelin effects on gonadotropin secretion in male and female rats. *Neurosci Lett.* 2004;362:103–107.
100. Rak-Mardyla A. Ghrelin role in hypothalamus-pituitary-ovarian axis. *J Physiol Pharmacol.* 2013;64:695–704.
101. Unniappan S, Peter RE. *Invitro* and *invivo* effects of ghrelin on luteinizing hormone and growth hormone release in goldfish. *Am J Physiol Regul Integr Comp Physiol.* 2004;286: 1093–1101.
102. Sokołowska-Mikołajczyk M, Socha M, Szczerbik P, Epler P. The effects of ghrelin on the *in vitro* spontaneous and GnRH-A stimulated luteinizing hormone (LH) release from the pituitary cells of common carp (Cyprinuscarpio L.). *Comp Biochem Physiol A.* 2009;153:386–390.
103. Furuta M, Funabashi T, Rimura F. Intracerebroventricular administration of ghrelin rapidly suppresses pulsatile luteinizing hormone secretion in ovariectomized rats. *Biochem Biophys Res Commun.* 2001;288:780–785.
104. Kawamura K, Sato N, Fukuda J, Kodama H, Kumegai J, Tanikawa H, Nakamura A, Honda Y, Sato T, Tanaka T. Ghrelin inhibits the development of mouse preimplantation embryos *in vitro.* *Endocrinology.* 2003;144:2623–2633.
105. Barreiro ML, Tena-Sempere M. Ghrelin and reproduction: a novel signal linking energy status and fertility? *Mol Cell Endocrinol.* 2004;226:1–9.

106. Vulliémoz NR, Xiao E, Xia-Zhang L, Germond M, Rivier J, Ferin M. Decrease in luteinizing hormone pulse frequency during a five-hour peripheral ghrelin infusion in the ovariectomized rhesus monkey. *J Clin Endocrinol Metab.* 2004;89:5718–5723.
107. Fernández-Fernández R, Tena-Sempere M, Navarro VM, Barreiro ML, Castellano JM, Aguilar E, Pinilla L. Effects of ghrelin upon gonadotropin-releasing hormone and gonadotropin secretion in adult female rats: *in vivo* and *in vitro* studies. *Neuroendocrinology.* 2005;82:245–255.
108. Iqbal J, Kurose Y, Canny B, Clarke IJ. Effects of central infusion of ghrelin on food intake and plasma levels of growth hormone, luteinizing hormone, prolactin, and cortisol secretion in sheep. *Endocrinology.* 2006;147:510–519.
109. Fernández-Fernández R, Martini AC, Navarro VM, Castellano JM, Dieguez C, Aguilar E, Pinilla L, Tena-Sempere M. Novel signals for the integration of energy balance and reproduction. *Mol Cell Endocrinol.* 2006;254–255:127–132.
110. Martini AC, Fernández-Fernández R, Tovar S, Navarro VM, Vigo E, Vazquez MJ, Davies JS, Thompson NM, Aguilar E, Pinilla L, Wells T, Dieguez C, Tena-Sempere M. Comparative analysis of the effects of ghrelin and unacylated ghrelin on luteinizing hormone secretion in male rats. *Endocrinology.* 2006;147:2374–2382.
111. Kluge M, Schüssler P, Uhr M, Yassouridis A, Steiger A. Ghrelin suppresses secretion of luteinizing hormone in humans. *J Clin Endocrinol Metab.* 2007;92:3202–3205.
112. Lanfranco F, Bonelli L, Baldi M, Me E, Broglio F, Ghigo E. Acylated ghrelin inhibits spontaneous LH pulsatility and responsiveness to naloxone, but not that to GnRH in young men: evidence for a central inhibitory action of ghrelin on the gonadal axis. *J Clin Endocrinol Metab.* 2008;93:3633–3639.
113. Muccioli G, Lorenzi T, Lorenzi M, Ghè C, Arnoletti E, Raso GM, Castellucci M, Gualillo O, Meli R. Beyond the metabolic role of ghrelin: a new player in the regulation of reproductive function. *Peptides.* 2011;32:2514–2521.
114. Kamegai J, Tamura H, Shimizu T, Ishii S, Sugihara H, Wakabayashi I. Central effect of ghrelin, an endogenous growth hormone secretagogue, on hypothalamic peptide gene expression. *Endocrinology.* 2000;141:4797–4800.
115. Forbes S, Li XF, Kinsey-Jones J, O'Byrne K. Effects of ghrelin on kisspeptin mRNA expression in the hypothalamic medial preoptic area and pulsatile luteinising hormone secretion in the female rat. *Neurosci Lett.* 2009;460:143–147.
116. Vanhorebeek I, Langouche L, Van den Berghe G. Endocrine aspects of acute and prolonged critical illness. *Nat Clin Pract Endocrinol Metab.* 2006;2:20–31.
117. Shimizu Y, Nagaya N, Isobe T, Imazu M, Okumura H, HosodaH. Kojima M, Kangawa K, Kohno N. Increased plasma ghrelin level in lung cancer cachexia. *Clin Cancer Res.* 2003;9:774–778.
118. Sirotkin AV, Chrenkovà M, Nitrayovà S, Patras P, Darlak K, Valenzuela F, Pinilla L, Tena-Sempere M. Effects of chronic food restriction and treatments with leptin or ghrelin on different reproductive parameters of mal rats. *Peptides.* 2008;29 (8):1362–1368.
119. Dupont J, Maillard V, Coyral-Castel S, Ramé C, Froment P. Ghrelin in female and male reproduction. *Int J Pept.* 2010;2010:1–8.
120. Ishikawa T, Fujioka H, Ishimura T, Takenaka A, Fujisawa M. Ghrelin expression in human testis and serum testosterone level. *J Androl.* 2007;28:320–324.
121. Repaci A, Gambineri A, Pagotto U, Pasquali R. Ghrelin and reproductive disorders. *Mol Cell Endocrinol.* 2011;340:70–79.
122. Roa J, Tena-Sempere M. Connecting metabolism and reproduction: roles of central energy sensors and key molecular mediators. *Mol Cell Endocrinol.* 2014;397:4–14.
123. Zigman JM, Elmquist JK. Minireview: from anorexia to obesity-the yin and yang of body weight control. *Endocrinology.* 2003;144:3749–3756.

124. Fernandez-Fernandez R, Navarro VM, Barreiro ML, Vigo EM, Tovar S, Sirotkin AV, Casanueva FF, Aguilar E, Dieguez C, Pinilla L, Tena-Sempere M. Effects of chronic hyperghrelinemia on puberty onset and pregnancy outcomein the rat. *Endocrinology.* 2005;146:3018–3025.
125. Lebrethon MC, Aganina A, Fournier M, Gérard A, Parent AS, Bourguignon JP. Effects of *in vivo* and *in vitro* administration of ghrelin, leptin and neuropeptide mediators on pulsatile gonadotrophin-releasing hormone secretion from male rat hypothalamus before and after puberty. *J Neuroendocrinol.* 2007;19:181–188.
126. El-Eshmawy MM, Abdel Aal IA, El Hawary AK. Association of ghrelin and leptin with reproductive hormones in constitutional delay of growth and puberty. *Reprod Biol Endocrinol.* 2010;8:153.
127. Garcia MC, Lopez M, Alvarez CV, Casanueva F, Tena-Sempere M, Dieguez C. Role of ghrelin in reproduction. *Reproduction.* 2007;133:531–540.
128. Gaytan F, Barreiro ML, Caminos JE, Chopin LK, Herington AC, Morales C, Pinilla L, Paniagua R, Nistal M, Casanueva FF, Aguilar E, Diéguez C, Tena-Sempere M. Expression of ghrelin and its functional receptor, the type 1a growth hormone secretagogue receptor, in normal human testis and testicular tumors. *J Clin Endocrinol Metab.* 2004;89:400–409.
129. Miller DW, Harrison JL, Brown YA, Doyle U, Lindsay A, Adam CL, Lea RG. Immunohistochemical evidence for an endocrine/paracrine role for ghrelin in the reproductive tissues of sheep. *Reprod Biol Endocrinol.* 2005;3:60.
130. Yan W, Kero J, Huhtaniemi I, Toppari J. Stem cell factor functions as a survival factor for mature Leydig cells and a growth factor for precursor Leydig cells after ethylene dimethanesulfonate treatment: implication of a role of the stem cell factor/c-Kit system in Leydig cell development. *Dev Biol.* 2000;227:169–182.
131. Barreiro ML, Gaytan F, Castellano JM, Suominen JS, Roa J, Gaytan M, Aguilar E, Dieguez C, Toppari J, Tena-Sempere M. Ghrelin inhibits the proliferative activity of immature Leydig cells *in vivo* and regulates stem cell factor messenger ribonucleic acid expression in rat testis. *Endocrinology.* 2004;145:4825–4834.
132. Zhang W, Lei Z, Su J, Chen S. Expression of ghrelin in the porcine hypothalamo-pituitary-ovary axis during the estrous cycle. *Anim Reprod Sci.* 2008;109:356–367.
133. Manning AJ, Murray HM, Gallant JW, Matsuoka MP, Radford E, Douglas SE. Ontogenetic and tissue-specific expression of preproghrelin in the Atlantic halibut, Hippoglossus hippoglossus L. *J Endocrinol.* 2008;196:181–192.
134. Gaytan F, Barreiro ML, Chopin LK, Herington AC, Morales C, Pinilla L, Casanueva FF, Aguilar E, Diéguez C, Tena-Sempere M. Immunolocalization of ghrelin and its functional receptor, the type 1a growth hormone secretagogue receptor, in the cyclic human ovary. *J Clin Endocrinol Metab.* 2003;88:879–887.
135. Viani I, Vottero A, Tassi F, Cremonini G, Sartori C, Bernasconi S, Ferrari B, Ghizzoni L. Ghrelin inhibits steroid biosynthesis by cultured granulosa-lutein cells. *J Clin Endocrinol Metab.* 2008;93:1476–1481.
136. Nekola MV, Coy DH. Direct and indirect inhibition of ovulation in rats by an antagonist of luteinizing hormone-releasing hormone. *Endocrinology.* 1985;116:756–760.
137. Lorenzi T, Meli R, Marzioni D, Morroni M, Baragli A, Castellucci M, Gualillo O, Muccioli G. Ghrelin: a metabolic signal affecting the reproductive system. *Cytokine Growth Factor Rev.* 2009;20:137–152.
138. Tropea A, Tiberi F, Minici F, Orlando M, Gangale MF, Romani F, Miceli F, Catino S, Mancuso S, Sanguinetti M, Lanzone A, Apa R. Ghrelin affects the release of luteolytic and luteotropic factors in human luteal cells. *J Clin Endocrinol Metab.* 2007;92:3239–3245.
139. Messini CI, Dafopoulos K, Chalvatzas N, Georgoulias P, Messinis IE. Growth hormone and prolactin response to ghrelin during the normal menstrual cycle. *Clin Endocrinol.* 2009;71:383–387.

140. Messini CI, Dafopoulos K, Malandri M, Georgoulias P, Anifandis G, Messinis IE. Inhibitory effect of submaximal doses of ghrelin on gonadotropin secretion in women. *Horm Metab Res.* 2014;46:36–40.
141. Brar AK, Frank GR, Kessler CA, Cedars MI, Handwerger S. Progesterone-dependent decidualization of the human endometrium is mediated by cAMP. *Endocrine.* 1997;6:301–307.
142. Feinman MA, Kliman HJ, Caltabiano S, Strauss III JF. 8-Bromo-3′,5′-adenosine monophosphate stimulates the endocrine activity of human cytotrophoblasts in culture. *J Clin Endocrinol Metab.* 1986;63:1211–1217.
143. Pansini F, Bergamini CM, Bettocchi Jr S, Malfaccini M, Santoiemma M, Scoppetta V, Bagni B, Mollica G. Sex steroid hormones influence the cAMP content in human endometrium during the menstrual cycle. *Gynecol Obstet Invest.* 1984;18:174–177.
144. Makino Y, Hosoda H, Shibata K, Makino I, Kojima M, Kangawa K, Kawarabayashi T. Alteration of plasma ghrelin levels associated with the blood pressure in pregnancy. *Hypertension.* 2002;39:781–784.
145. Fuglsang J, Skjaerbaek C, Espelund U, Frystyk J, Fisker S, Flyvbjerg A, Ovesen P. Ghrelin and its relationship to growth hormones during normal pregnancy. *Clin Endocrinol.* 2005;62:554–559.
146. Hehir MP, Glavey SV, Morrison JJ. Uterorelaxant effect of ghrelin on human myometrial contractility. *Am J Obstet Gynecol.* 2008;198:323.e1–323.e5.
147. Nakahara K, Nakagawa M, Baba Y, Sato M, Toshinai K, Date Y, Nakazato M, Kojima M, Miyazato M, Kaiya H, Hosoda H, Kangawa K, Murakami N. Maternal ghrelin plays an important role in rat fetal development during pregnancy. *Endocrinology.* 2006;147:1333–1342.
148. Cortelazzi D, Cappiello V, Morpurgo PS, Ronzoni S, Nobile De Santis MS, Cetin I, Beck-Peccoz P, Spada A. Circulating levels of ghrelin in human fetuses. *Eur J Endocrinol.* 2003;149:111–116.
149. Chanoine JP, Yeung LP, Wong AC, Birmingham CL. Immunoreactive ghrelin in human cord blood: relation to anthropometry, leptin, and growth hormone. *J Pediatr Gastroenterol Nutr.* 2002;35:282–286.
150. Farquhar J, Heiman M, Wong AC, Wach R, Chessex P, Chanoine JP. Elevated umbilical cord ghrelin concentrations in small for gestational age neonates. *J Clin Endocrinol Metab.* 2003;88:4324–4327.

CHAPTER TWO

Deconstructing the Growth Hormone Receptor (GHR): Physical and Metabolic Phenotypes of Tissue-Specific GHR Gene-Disrupted Mice

Jonathan A. Young[*,†], **Edward O. List**[*,‡], **John J. Kopchick**[*,§,1]

[*] Edison Biotechnology Institute, Ohio University, Athens, Ohio, USA
[†] Department of Biological Sciences, Ohio University, Athens, Ohio, USA
[‡] Department of Specialty Medicine, Heritage College of Osteopathic Medicine, Ohio University, Athens, Ohio, USA
[§] Department of Biomedical Sciences, Heritage College of Osteopathic Medicine, Ohio University, Athens, Ohio, USA
[1] Corresponding author: e-mail address: kopchick@ohio.edu.

Contents

Abstract

Growth hormone (GH)-induced signaling results in numerous effects in multiple tissues throughout the body. Elucidation of several of these effects has come from studies observing the various phenotypes of the GH receptor (GHR) gene-disrupted (GHR−/−) mouse. These mice are dwarf and obese with increased insulin sensitivity, are resistant to cancer and diabetes, and have an extended lifespan. While these mice have proven to be a valuable tool for understanding the pleiotropic effects of GH, we and others have generated novel tissue-specific GHR gene-disrupted mouse lines that are now helping to further dissect the actions of GH on specific cells/tissues. In this chapter, we summarize the various phenotypes observed in these mice.

Progress in Molecular Biology and Translational Science, Volume 138
ISSN 1877-1173
http://dx.doi.org/10.1016/bs.pmbts.2015.10.014

1. INTRODUCTION

Mice with deficiencies in growth hormone (GH) gene expression have been useful for uncovering some of the pleiotropic actions of this peptide hormone. Initially, dwarf mouse lines such as Ames and Snell were used. However, both of these mouse lines are a result of gene mutations (*Prop-1* and *Pit-1*, respectively), that are important for pituitary development. Phenotypic outcomes of these mutations result in multiple pituitary hormone deficiencies including: GH, prolactin, and thyroid-stimulating hormone. In order to obtain a mouse with GH action specifically removed, our laboratory generated the GH receptor (GHR) gene-disrupted mouse line (GHR−/−) in the mid 1990s.[1] These dwarf mice are completely resistant to GH action without disrupting other pituitary hormonal systems. Thus, GHR−/− mice quickly became a valuable tool in numerous mouse studies for understanding many of the physiologic effects of GH.[2,3]

Defining the direct effects of GH versus the indirect effects (via IGF-1) on a given tissue has been, and still is, challenging. According to the dual-effector theory of GH action proposed by Green and coworkers in the mid 1980s, GH exhibits direct actions on tissues by stimulating cellular differentiation (in the original theory) and also exhibits indirect actions via IGF-1-stimulated growth.[4] Since Green's original theory in 1985, many independent as well as intersecting actions for GH and IGF-1 have been described. For example, Efstratiadis' group in the early 2000s reported that GH and IGF-1 have both direct and overlapping effects on mouse growth.[5] A continuing quest to define direct versus indirect actions of GH remains challenging since GH and IGF-1 regulate one another, and as stated earlier, some of the actions of GH and IGF-1 are overlapping.

Our laboratory and others have generated tissue-specific GHR-disrupted mouse lines using the Cre/Lox system,[6,7] in order to better define the direct effects of GH on specific tissues. To date, five tissues/cell types have been targeted in seven distinct tissue-specific GHR gene-disrupted mouse lines.[8–21] In this review, we will give a brief account of several important findings in global GHR−/− mice. We will then summarize the findings from each of the tissue-specific GHRKO studies and compare these findings to results from global GHR−/− mice. Taken together, findings from tissue-specific GHR gene-disrupted mice compared to those from GHR−/− mice will help define the direct versus indirect actions of GH on specific tissues as well as generate novel and, sometimes, unexpected results. Later in the chapter, we first describe the GHR−/− mouse line and then the conditional/tissue-specific GHR gene-disrupted mice.

2. GLOBAL GHR GENE-DISRUPTED MICE

As stated earlier, our laboratory generated the GHR gene-disrupted mouse (GHR−/−) in the mid 1990s[22] to help elucidate the various *in vivo* actions of GH. Since GHR−/− mice are resistant (or insensitive) to GH, the phenotype of these mice provides a valuable "tool" to help define the various actions of GH. Accordingly, GHR−/− mice have been sent worldwide for studies of GH relative to aging, metabolism, apoptosis, reproduction, and other GH-mediated actions.[23] GHR−/− mice have a marked decrease in body size (see Fig. 1A) with greatly decreased circulating IGF-1 and elevated GH levels;[22] thus they are resistant or insensitive to GH. Organs are proportionally decreased in size with the exception of kidneys and liver, which are disproportionally smaller, while select adipose depots and brain are disproportionally larger.[24,25]

Besides the pronounced effects on body size, GHR−/− mice have numerous unique characteristics that highlight the various roles of GH on metabolism.[23] For instance, GHR−/− mice are remarkably insulin sensitive with low circulating insulin and low to normal levels of fasting blood glucose.[24,26–28] Additionally, GHR−/− mice are obese with elevated levels of leptin.[28–30] Also, both total and high molecular weight adiponectin levels are increased in GHR−/− mice.[31] Since adiponectin is usually negatively correlated with adiposity but positively correlated with improved insulin sensitivity, this increase in adiponectin appears to follow with the improved insulin sensitivity of GHR−/− mice. Another attribute of GHR−/− mice that might explain the healthy phenotype despite obesity is that the adiposity is increased primarily in the subcutaneous white adipose tissue (WAT) depot.[25,30] The notion that subcutaneous WAT is relatively healthier than visceral WAT depots is supported by several studies.[32] One of the most remarkable observations from GHR−/− mice is that they are extremely long-lived independent of sex and genetic background.[24,33–35] Contributing to the extended longevity found in these mice is the fact that they are also resistant to neoplastic diseases,[36] similar to Laron syndrome (LS) patients.[37–39] It will be interesting to establish whether this remarkable decrease in cancer is due to the lack of direct action of GH or a result of

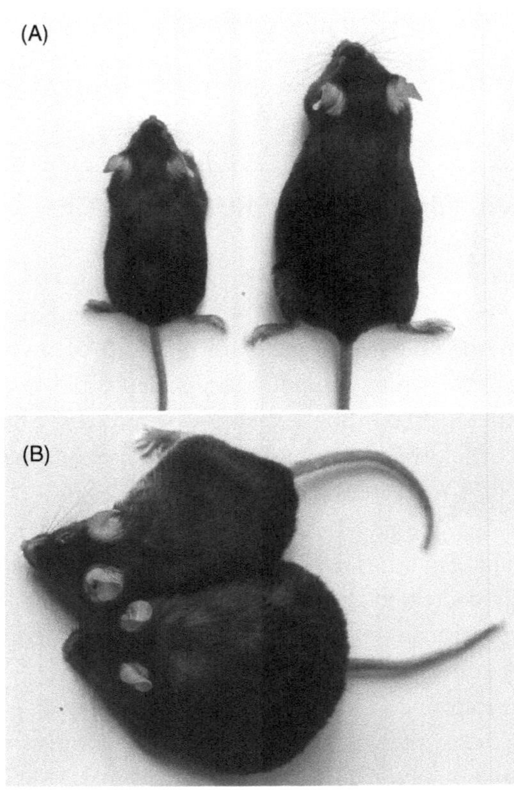

Figure 1 Images of global GHR−/− and fat-specific GHRKO mice. (A) 6-month-old global GHR−/− mouse (left) next to a littermate control (right). (B) 6-month-old FaGHRKO (bottom) and littermate control (top).

low total IGF-1 levels or both. Also, the mice are protected from high-fat diet-induced diabetes[26] and streptozotocin-induced glomerulosclerosis.[40] In this regard, a cohort of Ecuadorian LS patients have been found to be relatively resistant to the development of diabetes[38] although this has not been found in an Israeli cohort of LS patients.[3,41]

3. LIVER

One of the primary actions of GH is the stimulation of endocrine, paracrine, and autocrine IGF-1 in multiple tissues. For endocrine IGF-1, liver is the master organ since ~90% of circulating IGF-1 comes from this tissue in response to GH.[8,18] Since liver plays such an important role in the

GH/IGF-1 axis, liver-specific knockout of the GHR was a logical place to begin characterizing the differences between the actions of GH versus that of IGF-1. Since GH-induced intracellular signaling in the liver is necessary for IGF-1 production, disruption, or "knockout" of the GHR gene in the liver, which allows the direct and indirect (IGF-1 mediated) effects of GH to be differentiated from each other. Two separate lines of liver-specific GHR knockout mice have been reported, both using the albumin promoter/enhancer to drive Cre recombinase expression specific to the liver. Although the two lines are similar, they represent distinct populations of mice, and therefore have different names. GHRLD refers to the line reported by Fan *et al.* in 2009,[8] while LiGHRKO refers to the line reported by List *et al.* in 2014.[18]

Although the GHRLD and LiGHRKO lines were produced using the same techniques, their growth parameters differ. GHRLD mice show no change in body weight or composition,[8] while LiGHRKO mice are smaller and have increased adiposity early in life but decreased adiposity later in life (Table 1).[18] Since body composition changes are more pronounced later in life, these results don't necessarily disagree, as the GHRLD mice were measured at younger ages than the LiGHRKO. Both lines show increased liver weight and decreased kidney weight; LiGHRKO mice also show an increase in lung and heart weight and a decrease in spleen weight. Despite the differences in growth seen between the two lines, they share a similar

Table 1 Comparison of two liver-specific GHR gene disrupted mouse lines.

Mouse Line/ Cre Promoter	Body Metrics	Glucose Homeostasis	Tissue Sizes	Miscellaneous
GHRLD/ Albumin	No change in body weight or body composition	Glucose intolerant and insulin resistant	Increased liver size; decreased kidney size	Increased inflammation and fibrosis; decreased bone density
LiGHRKO/ Albumin	Decreased body weight and length; increased body fat at young ages and increased body weight at older ages	Increased blood glucose, increased insulin (males)	Increased liver, heart, and lung sizes; decreased kidney and spleen sizes	Increased grip strength, leptin (females), adiponectin, IL-6

Table 2 Comparison of two muscle-specific GHR gene disrupted mouse lines.

Mouse Line/ Cre Promoter	Body Metrics	Glucose Homeostasis	Muscle Parameters	Miscellaneous
ΔGHR/ mef-2c-73k	Increased weight Increased fat mass	Increased blood glucose Insulin resistant	Decreased grip strength Decreased muscle size	
mGHRKO/ MCK	Decreased weight Decreased lean mass Decreased fat mass	Insulin sensitive	Decreased muscle triglycerides	Increased adiponectin

metabolic phenotype. Impaired glucose homeostasis is common between both the lines, and males in both lines show increased liver steatosis, indicative of impaired lipid metabolism in the liver. An important finding concerning the LiGHRKO mice was that females do not have increased liver steatosis. Also, LiGHRKO mice have increased expression of adipokines, specifically resistin and adiponectin, with leptin increased only in females.[18] Also, GHRLD mice have increased fibrosis and inflammation;[8] these parameters have not been reported in LiGHRKO mice. LiGHRKO mice have increased grip strength, indicative of improved muscular health (Table 2).

Recall that GHR−/− mice, which have GHR knocked out globally, are dwarf and have increased adiposity throughout life. Thus, neither of the two liver-specific GHR knockout lines has the same growth characteristics as the global knockout. Interestingly, the adipose accumulation profile of LiGHRKO mice is similar to the profile seen in bovine GH transgenic (bGH) mice, which have excess GH-induced signaling throughout life. Finally, the liver-specific GHR knockout lines have impaired glucose homeostasis, in contrast to the enhanced glucose homeostasis of the GHR−/− mice.

As stated earlier, GHR−/− mice have a greatly increased lifespan, which is presumably due, in part, to their improved glucose homeostasis. However, the liver GHR knockout mice have a normal lifespan, which is counterintuitive with regards to their impaired glucose and insulin sensitivity. This suggests that other factors such as decreased circulating IGF-1 counteract the negative effects of impaired glucose homeostasis.

4. MUSCLE

Another tissue in which GH-induced signaling is metabolically important is the muscle. Systemic administration of GH increases muscle mass concurrently with the increase in IGF-1 production due to the endocrine GH action on the liver and/or the paracrine/autocrine action of GH in the muscle. As muscle is one of the prominent insulin-sensitive tissues, it is also of interest to study for further understanding the effects of GH on glucose homeostasis.

In order to distinguish between GH action and IGF-1 action on muscle tissue, muscle-specific GHR knockout mice were generated.[10,12–14] As with liver GHR knockout mice, two different muscle-specific GHR knockout mouse strains have been reported; however, the promoters/enhancers used in each differed. The first published results by Mavalli *et al.*,[10] used the *Mef-2c-73k* promoter/enhancer to drive Cre expression specifically in the muscle, while in the second published by Vijayakumar *et al.*,[12–14] Cre recombinase expression was driven by the muscle creatine kinase (MCK) promoter/enhancer. The *Mef-2c-73k* promoter is specific to muscle in the adult mouse, but is expressed additionally in the heart and brain during fetal life, while the MCK promoter is specific to muscle (both skeletal and cardiac) in all stages of development. Therefore, it is important to keep in mind that any differences between the different muscle-specific GHR knockout lines may be due to differences in the pattern of Cre expression.

Possibly due to the use of different promoters/enhancers, the changes in growth and body composition in these two lines of mice contradict each other, with the *Mef* mice showing an increase in body weight and adiposity, while the MCK mice show decreases in both, as well as a decrease in lean mass. Also, glucose homeostasis is impaired in the *Mef* mice, while the MCK mice have increased insulin sensitivity.

Finally, the muscle structure differs in *Mef* mice, with decreased myofiber size and myonuclei number, as well as a transition from type I fibers to type II fibers. As a result of these structural changes, the mice have decreased muscle strength and endurance, as measured by grip strength and rotarod tests, respectively. In contrast, the MCK mice show no structural changes in the muscle, but show altered lipid metabolism as well as decreased inflammation and increased adiponectin. The differences seen between the two models of muscle-specific GHR knockout mice underscore the importance of a carefully selected promoter/enhancer when using the Cre-lox system to generate gene disrupted or knockout animals. Unfortunately, as the majority of published work on the muscle-specific GHR knockout animals only includes male mice, it is difficult to compare these results with the results from the liver-specific GHR knockout animals that show a phenotype that is strongly dependent on sex.

5. ADIPOSE

One of the most striking characteristics of the global GHR knockout mice is their long lifespan despite having excess fat: a state of "healthy obesity". Therefore, adipose-specific knockout of GHR gene should help in understanding the healthy obese phenotype of the global knockout animals. Furthermore, adipose tissue is another prominent insulin-sensitive tissue; thus GH action on adipose tissue could contribute to the effects of GH on glucose homeostasis.

We have used the Fabp4 (also called aP2) promoter/enhancer to drive Cre recombinase expression, generating the adipose-specific GHR knockout (FaGHRKO) mice.[16] These mice have increased body weight compared to controls, primarily due to increased fat mass (see Fig. 1B). Although the increase in fat mass in FaGHRKO mice aligns with the increase in fat mass in the global knockout, the improvement in glucose homeostasis seen in the global knockout is not seen in the FaGHRKO. This indicates that the fat mass present does not have a large effect on glucose homeostasis in this mouse line. Although fat mass is increased in FaGHRKO mice as well as GHR−/−, there are differences in the depot or depots where the additional fat is accumulating. FaGHRKO mice have increased fat mass in all fat depots, including the subcutaneous, retroperitoneal, and mesenteric, while GHR−/− mice see an increase specifically in the subcutaneous depot. The difference in deposition of excess fat is one possible explanation for the differences in glucose homeostasis seen in these mice.

FaGHRKO mice, like LiGHRKO mice, show many sex-specific changes. First of these is the increased rate at which the male FaGHRKO mice gain weight compared to the females, with males showing significant changes 2 months earlier. Male FaGHRKO mice also show an increase in IGF-1 levels, as well as decreased spleen mass and total adiponectin as compared to controls. These changes were not seen in the female FaGHRKO mice; however, female animals reveal an increase in soleus muscle and kidney mass and leptin, IGFBP5, and IL-6 levels, along with decreased lean mass.

The multitude of differences seen between males and females underscores the importance of measuring both sexes when studying the GH/IGF-1 axis. Finally, despite the normal glucose homeostasis seen in the fat-specific knockout, these mice show a decrease in lifespan (List and Kopchick, unpublished data). This suggests that disruption of GHR specifically in adipose tissue does not promote "healthy obesity" and indicates that factors other than glucose homeostasis are responsible for determining lifespan in this case.

6. PANCREATIC BETA (β) CELL

The changes in glucose homeostasis seen in GHRKO−/− and various tissue-specific knockout strains underscores the importance of GH in insulin action; thus it is possible that GH has a direct effect on the β cells in the pancreas that produce insulin. Indeed, GH treatment has been shown to cause proliferation of β cells.[42] To further examine the direct effects of GH on the β cells, Wu *et al.* generated a pancreatic β-cell-specific GHR knockout mouse (βGHRKO).[43] Glucose homeostasis of these mice was then tested on a high-fat diet in addition to the standard chow diet. The βGHRKO mice experienced an impaired insulin response on the high-fat diet, due to a decrease in β-cell mass. This change was not seen on the normal chow diet. The results of this study indicate that direct GH action on β cells improves glucose homeostasis through increasing cell proliferation. Although GH is a diabetogenic molecule, this study shows that GH's direct effects on the pancreas may actually be protective against diabetes. The necessity of stressing the mouse with a high-fat diet in order to see the benefit could explain why the negative effect of knocking out GHR in the β cells is not seen in the GHR−/− mice.

7. MACROPHAGE

A recent publication,[44] as well as results from the mGHRKO[12] and GHRLD[21] mice, link GH-induced intracellular signaling with alterations in the inflammatory state. Since macrophages are one of the major mediators of inflammation in GH-sensitive tissues, generating a macrophage-specific GHR knockout mouse would allow the direct effects of GH on the inflammatory response to be better understood. Lu *et al.* generated a

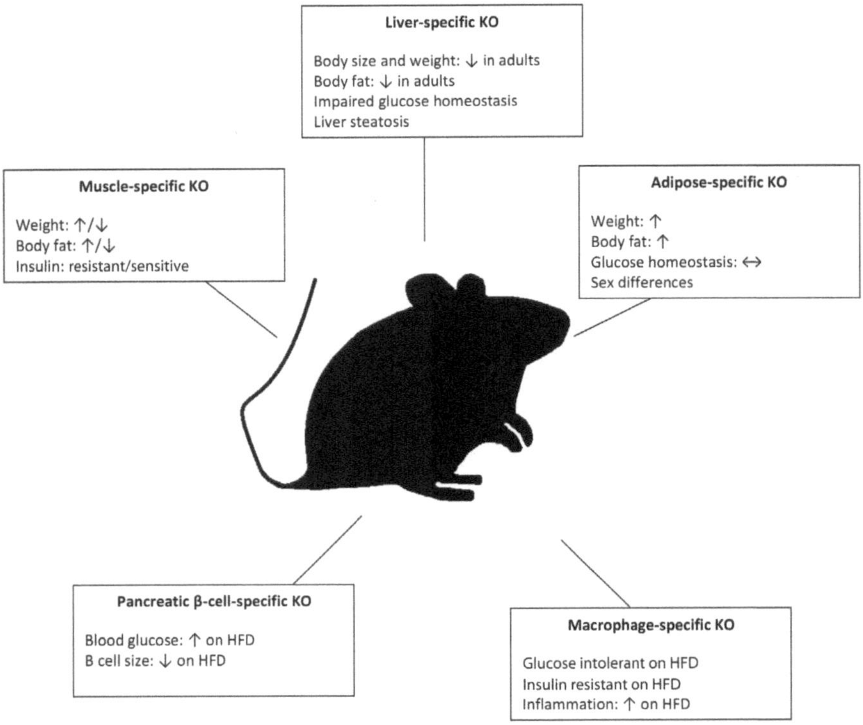

Figure 2 Overview of the main phenotypes found in tissue-specific GHRKO mice.

macrophage-specific GHR knockout (MacGHRKO) mouse using the LysM promoter/enhancer to direct Cre expression.[15] MacGHRKO mice were then fed a high-fat diet and their body composition, glucose homeostasis, and inflammatory state were determined. Although the body composition of the MacGHRKO mice was similar to the controls, MacGHRKO mice saw an increase in the size of the epididymal adipose depot and an increase in adipocyte size. The MacGHRKO mice also developed glucose intolerance and insulin resistance, marked by a decrease in insulin-induced Akt phosphorylation specific to adipose tissue. Interestingly, the FaGHRKO mice have normal glucose homeostasis, indicating that macrophages may be the major contributor to glucose sensitivity in the adipose tissue. MacGHRKO mice also saw increased adipose inflammation on the high-fat diet, indicated by increased M1 (proinflammatory) macrophages and decreased M2 (anti-inflammatory) macrophages in epididymal adipose tissue.

8. CONCLUSIONS

The phenotype of the GHR−/− mouse is a result of organism-wide disruption of GH-induced signaling, which results in major changes in the growth, metabolism, and longevity of the mice. Although GHR has been specifically disrupted in many tissues and cell types, (Fig. 2) including the major GH- and insulin-responsive tissues, the phenotype of the global GHR−/− mouse was not replicated in any of the tissue-specific gene-disrupted mice. This indicates that the phenotype of the GHR−/− mouse is a result of multiple tissue changes in GH action. Future results involving crossing of the various conditional GHR−/− mice or establishing other GHR−/− line may help resolve this puzzle.

REFERENCES

1. Zhou Y, He L, Baumann G, Kopchick JJ. Deletion of the mouse GH-binding protein (mGHBP) mRNA polyadenylation and splicing sites does not abolish production of mGHBP. *J Mol Endocrinol.* 1997;19(1):1–13.
2. List EO, Sackmann-Sala L, Berryman DE, et al. Endocrine parameters and phenotypes of the growth hormone receptor gene disrupted (GHR−/−) mouse. *Endocr Rev.* 2011;32(3):356–386.
3. Laron Z, Kopchick J. *Laron Syndrome—From Man to Mouse Lessons From Clinical and Experimental Experience.* Berlin; New York: Springer; 2011.
4. Green H, Morikawa M, Nixon T. A dual effector theory of growth-hormone action. *Differentiation.* 1985;29(3):195–198.
5. Lupu F, Terwilliger JD, Lee K, Segre GV, Efstratiadis A. Roles of growth hormone and insulin-like growth factor 1 in mouse postnatal growth. *Dev Biol.* 2001;229(1):141–162.
6. Sauer B. Inducible gene targeting in mice using the Cre/lox system. *Methods.* 1998;14(4):381–392.
7. Sauer B, Henderson N. Site-specific DNA recombination in mammalian cells by the Cre recombinase of bacteriophage P1. *Proc Natl Acad Sci USA.* 1988;85(14):5166–5170.
8. Fan Y, Menon RK, Cohen P, et al. Liver-specific deletion of the growth hormone receptor reveals essential role of GH signaling in hepatic lipid metabolism. *J Biol Chem.* 2009;284:19937–19944.
9. Lu C, Kumar PA, Fan Y, Sperling MA, Menon RK. A novel effect of growth hormone on macrophage modulates macrophage-dependent adipocyte differentiation. *Endocrinology.* 2010;151(5):2189–2199.
10. Mavalli MD, DiGirolamo DJ, Fan Y, et al. Distinct growth hormone receptor signaling modes regulate skeletal muscle development and insulin sensitivity in mice. *J Clin Invest.* 2010;120(11):4007–4020.
11. Wu Y, Liu C, Sun H, et al. Growth hormone receptor regulates beta cell hyperplasia and glucose-stimulated insulin secretion in obese mice. *J Clin Invest.* 2011;121(6):2422–2426.
12. Vijayakumar A, Wu Y, Sun H, et al. Targeted loss of GHR signaling in mouse skeletal muscle protects against high-fat diet-induced metabolic deterioration. *Diabetes.* 2012;61(1):94–103.

13. Vijayakumar A, Wu Y, Buffin NJ, et al. Skeletal muscle growth hormone receptor signaling regulates basal, but not fasting-induced, lipid oxidation. *PLoS One.* 2012;7(9): e44777.
14. Vijayakumar A, Buffin NJ, Gallagher EJ, et al. Deletion of growth hormone receptors in postnatal skeletal muscle of male mice does not alter muscle mass and response to pathological injury. *Endocrinology.* 2013;154(10):3776–3783.
15. Lu C, Kumar PA, Sun J, et al. Targeted deletion of growth hormone (GH) receptor in macrophage reveals novel osteopontin-mediated effects of GH on glucose homeostasis and insulin sensitivity in diet-induced obesity. *J Biol Chem.* 2013;288(22):15725–15735.
16. List EO, Berryman DE, Funk K, et al. The role of GH in adipose tissue: lessons from adipose-specific GH receptor gene-disrupted mice. *Mol Endocrinol.* 2013;27(3):524–535.
17. Li X, Bartke A, Berryman DE, et al. Direct and indirect effects of growth hormone receptor ablation on liver expression of xenobiotic metabolizing genes. *Am J Physiol Endocrinol Metab.* 2013;305:E942–E950.
18. List EO, Berryman DE, Funk K, et al. Liver-specific GH receptor gene-disrupted (LiGHRKO) mice have decreased endocrine IGF-I, increased local IGF-I, and altered body size, body composition, and adipokine profiles. *Endocrinology.* 2014;155(5):1793–1805.
19. Gesing A, Wang F, List EO, et al. Expression of apoptosis-related genes in liver-specific growth hormone receptor gene-disrupted mice is sex dependent. *J Gerontol A Biol Sci Med Sci.* 2015;70:44–52.
20. Dominick G, Berryman DE, List EO, et al. Regulation of mTOR activity in Snell dwarf and growth hormone receptor gene-disrupted mice. *Endocrinology.* 2014;156:565–575.
21. Fan Y, Fang X, Tajima A, et al. Evolution of hepatic steatosis to fibrosis and adenoma formation in liver-specific growth hormone receptor knockout mice. *Front Endocrinol (Lausanne).* 2014;5:218.
22. Zhou Y, Xu BC, Maheshwari HG, et al. A mammalian model for Laron syndrome produced by targeted disruption of the mouse growth hormone receptor/binding protein gene (the Laron mouse). *Proc Natl Acad Sci USA.* 1997;94(24):13215–13220.
23. List EO, Sackmann-Sala L, Berryman DE, et al. Endocrine parameters and phenotypes of the growth hormone receptor gene disrupted (GHR−/−) mouse. *Endocr Rev.* 2011;32(3): 356–386.
24. Coschigano KT, Holland AN, Riders ME, List EO, Flyvbjerg A, Kopchick JJ. Deletion, but not antagonism, of the mouse growth hormone receptor results in severely decreased body weights, insulin and IGF-1 levels and increased lifespan. *Endocrinology.* 2003;144(9): 3799–3810.
25. Berryman DE, List EO, Palmer AJ, et al. Two-year body composition analyses of long-lived GHR null mice. *J Gerontol A Biol Sci Med Sci.* 2010;65(1):31–40.
26. Berryman DE, List EO, Kohn DT, Coschigano KT, Seeley RJ, Kopchick JJ. Effect of growth hormone on susceptibility to diet-induced obesity. *Endocrinology.* 2006;147(6): 2801–2808.
27. Liu JL, Coschigano KT, Robertson K, et al. Disruption of growth hormone receptor gene causes diminished pancreatic islet size and increased insulin sensitivity in mice. *Am J Physiol Endocrinol Metab.* 2004;287(3):E405–E413.
28. Al-Regaiey KA, Masternak MM, Bonkowski M, Sun L, Bartke A. Long-lived growth hormone receptor knockout mice: interaction of reduced insulin-like growth factor i/insulin signaling and caloric restriction. *Endocrinology.* 2005;146(2):851–860.
29. Egecioglu E, Bjursell M, Ljungberg A, et al. Growth hormone receptor deficiency results in blunted ghrelin feeding response, obesity, and hypolipidemia in mice. *Am J Physiol Endocrinol Metab.* 2006;290(2):E317–E325.
30. Berryman DE, List EO, Coschigano KT, Behar K, Kim JK, Kopchick JJ. Comparing adiposity profiles in three mouse models with altered GH signaling. *Growth Horm IGF Res.* 2004;14(4):309–318.

31. Lubbers ER, List EO, Jara A, et al. Adiponectin in mice with altered GH action: links to insulin sensitivity and longevity? *J Endocrinol.* 2013;216(3):363–374.
32. Lee MJ, Wu Y, Fried SK. Adipose tissue heterogeneity: implication of depot differences in adipose tissue for obesity complications. *Mol Aspects Med.* 2013;34(1):1–11.
33. Coschigano KT, Clemmons D, Bellush LL, Kopchick JJ. Assessment of growth parameters and life span of GHR/BP gene-disrupted mice. *Endocrinology.* 2000;141(7): 2608–2613.
34. Bartke A, Brown-Borg H. Life extension in the dwarf mouse. *Curr Top Dev Biol.* 2004;63:189–225.
35. Bartke A, Coschigano K, Kopchick J, et al. Genes that prolong life: relationships of growth hormone and growth to aging and life span. *J Gerontol A Biol Sci Med Sci.* 2001;56(8): B340–B349.
36. Ikeno Y, Hubbard GB, Lee S, et al. Reduced incidence and delayed occurrence of fatal neoplastic diseases in growth hormone receptor/binding protein knockout mice. *J Gerontol A Biol Sci Med Sci.* 2009;64(5):522–529.
37. Laron Z, Pertzelan A, Mannheimer S. Genetic pituitary dwarfism with high serum concentration of growth hormone—a new inborn error of metabolism? *Isr J Med Sci.* 1966;2(2):152–155.
38. Guevara-Aguirre J, Balasubramanian P, Guevara-Aguirre M, et al. Growth hormone receptor deficiency is associated with a major reduction in pro-aging signaling, cancer, and diabetes in humans. *Sci Transl Med.* 2011;3(70):70ra13.
39. Shevah O, Laron Z. Patients with congenital deficiency of IGF-I seem protected from the development of malignancies: a preliminary report. *Growth Horm IGF Res.* 2007;17(1): 54–57.
40. Bellush LL, Doublier S, Holland AN, Striker LJ, Striker GE, Kopchick JJ. Protection against diabetes-induced nephropathy in growth hormone receptor/binding protein gene-disrupted mice. *Endocrinology.* 2000;141(1):163–168.
41. Laron Z, Avitzur Y, Klinger B. Insulin resistance in Laron syndrome (primary insulin-growth factor-I [IGF-I] deficiency) and effect of IGF-I replacement therapy. *J Pediatr Endocrinol Metab.* 1997;10:105–115.
42. Nielsen JH, Svensson C, Galsgaard ED, Moldrup A, Billestrup N. Beta cell proliferation and growth factors. *J Mol Med.* 1999;77(1):62–66.
43. Wu Y, Liu C, Sun H, et al. Growth hormone receptor regulates beta cell hyperplasia and glucose-stimulated insulin secretion in obese mice. *J Clin Invest.* 2011;121(6):2422–2426.
44. Benencia F, Harshman S, Duran-Ortiz S, et al. Male bovine GH transgenic mice have decreased adiposity with an adipose depot-specific increase in immune cell populations. *Endocrinology.* 2015;156:1794–1803.

CHAPTER THREE

Regulation of Growth Hormone by the Splanchnic Area

Silvia Barja-Fernandez*[,†,‡], Cintia Folgueira*[,‡], Cecilia Castelao*[,‡], Rosaura Leis[†], Ana B. Crujeiras[‡,§], Felipe F. Casanueva[‡,§], Luisa M. Seoane*[,†,1]

[*] Grupo Fisiopatología Endocrina, Instituto de Investigacion Sanitaria de Santiago de Compostela (IDIS), Complexo Hospitalario Universitario de Santiago (CHUS/SERGAS), Spain
[†] Pediatric Department, Universidade de Santiago de Compostela, Instituto de Investigacion Sanitaria de Santiago de Compostela (IDIS), Complexo Hospitalario Universitario de Santiago (CHUS/SERGAS), Spain
[‡] CIBER Fisiopatología Obesidad y Nutrición, Instituto de Salud Carlos III, Spain
[§] Laboratorio de Endocrinología Molecular y Celular, Instituto de Investigación Sanitaria de Santiago de Compostela (IDIS), Complexo Hospitalario Universitario de Santiago (CHUS/SERGAS), Santiago de Compostela, Spain
[1] Corresponding author: e-mail address: luisamaria.seoane@usc.es/Luisa.Maria.Seoane.Camino@sergas.es.

Contents

Abstract

The regulation of growth hormone (GH) was traditionally thought to be under the control of two main hypothalamic neuropeptides; GH-releasing hormone and somatostatin. In 1999, with the isolation of ghrelin, as a gastric-derived peptide with potent GH-releasing activity, concept of regulation of the somatotropic axis completely changed. In

Progress in Molecular Biology and Translational Science, Volume 138
ISSN 1877-1173
http://dx.doi.org/10.1016/bs.pmbts.2015.10.007

addition to its GH-releasing activity, ghrelin exhibited the capacity to modulate food intake and body weight. The role of this splanchnic factor in regulating GH as a nexus of energy balance control and GH are explored in this chapter. From a physiological standpoint, a novel mechanism of GH regulation mediated by ghrelin exists, implicating the peripheral modulation of the cannabinoid receptor.

1. GH

Growth hormone (GH) is produced by somatotroph cells located in the anterior pituitary gland and constitutes 25% of the total hormonal pituitary production.[1] The most common form of GH in plasma has a molecular weight of 22 kDa, although other forms weighing 20, 17.5, and 5 kDa are expressed at very low levels.[1–3] In humans, the GH gene activity is controlled by several factors, such as thyroid hormones, which inhibit GH gene transcription; glucocorticoids, which increase GH transcription; and pituitary-specific transcription factor-1 (Pit-1), which acts on the somatotroph cells of the anterior pituitary to stimulate GH gene expression.[1]

The receptor for GH (GHR) is a transmembrane receptor that belongs to the superfamily of class 1 cytokine proteins, and binding of GH to this receptor leads to the GHR dimerization. GH activation of growth hormone secretagogue receptor (GHSR) dimers induces the phosphorylation of Janus kinase 2 (JAK2). The main signaling pathway for GHR includes the signal transducer and activator of transcription (STAT) family of transcription factors, especially STAT5, which is the principal transcriptional effector phosphorylated by JAK2.[4] Some main biological effects of GH, such as IGF-1 induction, are mediated by this pathway.[5] In addition, other GH signal transduction pathways include the RAS/MAP kinase and phosphatidylinositol 3́kinase (PI3K/Akt) pathways.[6] Several evidences have shown that SOCS is also involved in the responses to GH.[1]

1.1 GH Effects

The activation of GHR by GH binding induces a wide range of pleiotropic effects, as the stimulatory action of protein synthesis acts directly on the cartilage and other tissues, thereby stimulating bone growth and increasing cell numbers at this level.[7] The major GH effects are mediated by IGF-1, which is produced in the liver by GH. Accordingly, GH is involved in longitudinal growth in the postnatal period, and its hypersecretion induces gigantism and acromegaly in adults.[8] The metabolic effects of GH include the regulation of carbohydrate and lipid metabolism in muscle and adipose tissue.[9–11] GH induces lipolysis at the adipose level, through the modulation of hormone-sensitive lipase activity, which is mediated by the activation of β-adrenergic receptors.[12] In addition, it has been described as an effect of GH, in decreasing triglyceride accumulation in adipose tissue, mediated by the inhibition of lipoprotein lipase.[13,14] The cell death-inducing DFF45-like effector (CIDEA) has also been proposed to be involved in the GH's action in the modulation of lipid droplets.[15] GH was suggested to affect brown adipose tissue metabolism through the modulation of uncoupling protein 1 and its interaction with CIDEA.[16] By supporting the role of GH in lipid metabolism regulation, changes in GH levels have been associated with alterations in the adipokine profile, including elevated adiponectin levels during GH deficiency.

GH is also involved in glucose homeostasis. In the liver, it was shown that GH stimulates hepatic glucose production. However, its relationship with the uptake or storage of liver glucose is unclear.[12] In the pancreas, it was shown that the GHR is directly involved in the response of β-cells to a high-fat diet, modulating insulin and glucose metabolism at this level.[17]

1.2 GH Regulation

GH is secreted with a pulsatile pattern in all the species, and is shown to get affected by multiple factors, such as nutrition, those related to body composition, metabolism, sexual steroids, thyroid hormones, as well as hepatic and renal functions.

The pulsatile secretion of GH was classically assumed to be under the control of two main hypothalamic neuropeptides namely, growth hormone releasing hormone (GHRH) and somatostatin (SS). In addition, IGF-1 is also involved in the regulation of GH secretion by inhibiting GH secretion through a feedback mechanism that stimulates SS and inhibits GHRH release from the hypothalamus. In addition, it was shown that IGF-1 has a direct effect on GH secretion in the somatotroph cells, independent of GHRH and SS.[18,19]

However, several studies have suggested that an additional factor independent of GHRH and SS is also involved in GH regulation.

The field of GH research advanced significantly with the isolation of ghrelin. Previously, the development of the peptidergic and non-peptidergic GH secretagogues with a potent effect on the liberation of GH,[20] and the isolation of the endogenous receptor for these GHSR, led the scientific community to believe that an endogenous peptide that bound to the GHSR constituted a novel pathway to regulate GH secretion.

2. GHRELIN

In 1999, ghrelin was isolated from the stomach as an endogenous ligand for the GHS orphan receptor, opening up a new field in hormone research.[21]

The discovery of ghrelin was the first clear evidence of a GH regulation by the splanchnic area. The main source of ghrelin production is the stomach, as shown by the fact that circulating ghrelin levels decrease by 65% after gastrectomy in humans,[22] and by 80% in rodents.[23] Ghrelin-producing cells in the gastrointestinal tract are classified into two types namely: the open-type cells of the stomach, which are regulated by luminal information, and the close-type cells of the gut, which are regulated by hormonal and neural stimulation.

Ghrelin is a 28-amino acid peptide produced from its 117-amino acid precursor preproghrelin.[21] The enzyme responsible for preproghrelin processing in the stomach is prohormone convertase 1/3 (PC1/3). The third Ser residue of ghrelin undergoes n-octanoylation, which is necessary for binding to the GHSR 1a. A total of 30% of circulating ghrelin is acetylated by n-octanoic acid at Ser3. The enzyme responsible for ghrelin octanoylation is ghrelin O-acyltransferase (GOAT), which was first identified in 2008.[24] This enzyme is expressed in the oxyntic mucosa of the stomach and colocalizes with ghrelin.[25] In the stomach, GOAT allows the acylation of ghrelin to produce the active form, which binds GHSR 1a to induce food intake, adiposity, and body weight gain. Indeed, GOAT knockout mice do not have the octanoylated ghrelin and show increased levels of des-acyl ghrelin compared to wild-type mice. It has been found that GOAT is regulated by nutritional status,[26] and the levels of the enzyme are highly influenced by the nature of the lipids in the diet.[27]

2.1 Main Factors Affecting Ghrelin Regulation

2.1.1 Nutritional Status

The regulation of ghrelin production by the factors related to the control of energy balance, has been the focus on a wide range of works published over the last 15 years. It was known that nutritional status is a key factor, which modulates ghrelin production in the stomach. Circulating levels of ghrelin vary depending on the nutritional status. They are increased under fasting conditions and recover to basal levels, when refed after a fasting period. Furthermore, ghrelin itself is able to regulate food intake-related behavior.[25,28]

The gastric mechanism, by which ghrelin regulates food intake was elucidated by findings, showing that plasma ghrelin regulation has its initial point of regulation at the gastric tissue level. It was shown that changes in the nutritional status first affect ghrelin expression and secretion at the gastric level, which is thereafter reflected in the levels of circulating ghrelin.[29] Moreover, it has been shown that the exposure to food-related sensory stimuli without subsequent food intake, modifies ghrelin secretion by the stomach and the circulating levels of ghrelin in the same way as true feeding.[29] This fact suggests that factors other than those produced by nutrient ingestion or absorption are related to ghrelin regulation.[30] Additionally, in 2007 Seoane *et al.* reported that ghrelin secretion from the stomach is influenced by food-related stimuli through a medium-term memory mechanism in the sensory neural system integrated into the enteric nervous system. Vagus nerve is thought to mediate the main effects of ghrelin, as the regulation of gastric ghrelin secretion by sensory stimuli gets blocked after surgical vagotomy in rats.[29]

2.1.2 GH Axis Components

The link between gastrointestinal tract hormone production and GH axis was evidenced by the findings reflecting that, in addition to food intake-related factors, other hormonal factors, such as the classical components of somatotropic axis, are implicated in gastric ghrelin regulation. Role of the components of the somatotropic axis on gastric ghrelin secretion were tested using an organ culture model of gastric tissue, which is capable of assessing the direct regulation of ghrelin secretion by the stomach. The *in vitro* incubation of gastric explants with SS, significantly decreased direct ghrelin secretion from the stomach.[31] These findings are in accordance with previous data in the literature, suggesting that some SS-producing cells have direct cellular contact with ghrelin-producing cells in the gastric funds.[32] Moreover, SS receptor expression has been described in the gastric mucosa.[33] In this context, the reduction in plasma ghrelin levels by SS is mediated by the direct inhibitory action of SS on the stomach.[34] Several studies in the literature suggest the existence of negative GH feedback on stomach ghrelin.[35] In addition, GHRs are present in the stomach and intestine.[36] By using the organ explant model, the first evidence of a direct GH effect on gastric ghrelin secretion was proposed, based on the fact that direct GH treatment on gastric tissue explants induced an inhibitory effect on ghrelin release from the stomach.[31] Contrarily, the authors did not find any effect of GHRH and IGF-1 on basal gastric ghrelin secretion.

3. GHRELIN AS A GHS

The discovery of ghrelin as the natural ligand for the GHSR,[21] presented a new perspective on the regulation of GH–neuroendocrine axis. In fact, ghrelin is mainly secreted by the stomach and is the only known peripheral orexigenic hormone, which induces GH release when exogenously administered at both central (i.c.v.) and peripheral (i.p.) levels. However, the entire mechanism responsible for this effect has not been completely elucidated.

In 1999, Kojima *et al.* showed that ghrelin induces GH release *in vitro* when administered to primary cultured pituitary cells with a level of potency comparable to that of GHRH. In the organism, ghrelin is produced by the stomach and secreted into the circulation, which induces GH production by the somatotroph cells.[21] It was initially shown that ghrelin specifically stimulates GH production because the secretory effects of administered ghrelin on other pituitary hormones, such as ACTH, FSH, LH, PRL, and TSH, were not observed, even at very high doses. As opposed to its *in vitro* actions, it was shown that ghrelin exerts other effects, such as the stimulation of the secretion of prolactin and ACTH secretion, and the inhibition of the gonadal axis.[37–39] The hypothalamus might be a key organ involved in ghrelin-induced GH secretion because it has been shown *in vitro* that the GH-releasing effect of GHS is higher in the hypothalamic–pituitary systems than in pituitary cells,[40] which is supported by the fact that more potent GH-releasing effects of ghrelin are found *in vivo* than *in vitro*.[41,42] Moreover, it was found that GHS has lesser effect on GH-releasing activity in animals with lesions in the pituitary stalk,[37] and human patients with traumatic brain injuries.[43]

Seoane *et al.* were the first to show that ghrelin induces a relevant increase in plasma GH at all the doses in freely moving rats. This stimulatory effect of ghrelin on GH is two- to threefold of that observed with GHRH, at the same doses of AUC and during maximal response terms.[44]

3.1 The Main Mechanism Involved in Ghrelin's Effects on GH Secretion

It is known that, although ghrelin and GHRH activate different receptors, the GHRH system should be operative in order to induce the GH-releasing activity of ghrelin.[20] In addition, GHRH was shown to increase ghrelin[45] expression at the pituitary level. Very little data exist on the ghrelin–SS interaction. However, *in vitro* studies have shown that ghrelin inhibits stimulated

SS release with no effect on basal SS production.[46] In 2003, Tannembaun's group proposed that ghrelin acts as an antagonist of SS. However, at the pituitary level, they found that the GH-releasing activity of ghrelin, is independent of the inhibition of endogenous SS release, as the SS-expressing cells in the hypothalamus are not direct targets of ghrelin action.[47]

In situ studies showed that the administration of ghrelin to rats did not affect GHRH or SS mRNA in the arcuate or paraventricular nucleus, respectively. This was probably due to the feedback induced by GH itself, which supposedly masked the effects of ghrelin on neuropeptides.[48] In dwarf rats, it was shown that the GHRH mRNA in the arcuate nucleus remained constant after ghrelin administration, but SS expression in the paraventricular nucleus was increased.

3.2 Central Nervous System Implications in Ghrelin's Regulation of GH

Ghrelin was shown to induce GH secretion in a more potent fashion than GHRH. It has been postulated that exogenously administered ghrelin induces GH production by two different mechanisms namely: ghrelin induces GH release by reaching the pituitary through the circulation, implicating a direct interaction with the GHSR in the somatotroph cells;[21,49] and ghrelin also exerts its actions through the autonomous nervous system (ANS) by the neuronal pathway through the vagus nerve.[50]

It was reported that the vagus nerve is required to induce ghrelin's maximum effect on GH secretion.[40] Moreover, it was described that after vagal blockade, a reduction in GH secretion by ghrelin was induced.[50] Al-Massadi *et al.* found that vagal disruption after surgical vagotomy induced a decrease in basal GH levels, reduced IFG-1 levels and a decreased GH's response to peripheral ghrelin injection.[51] Moreover, it was also found that GH's response to centrally administered ghrelin was also attenuated after surgical vagotomy, which clearly indicates that the peripheral and central pathways involved in the regulation of GH by ghrelin are connected through the vagus nerve (Fig. 1).[51]

Based on these findings, a novel mechanism behind ghrelin's peripheral effect on GH secretion mediated by the vagus nerve was fully described. It was shown that the vagus nerve is crucial to peripheral ghrelin's GH-releasing action at central level, as shown by the fact that when vagal signaling is disrupted, GHRH expression in the hypothalamic arcuate nucleus gets downregulated.[51] Furthermore, downregulation of GHRH and GHSRs in the pituitary, was also observed in animals with a disrupted vagal pathway. All

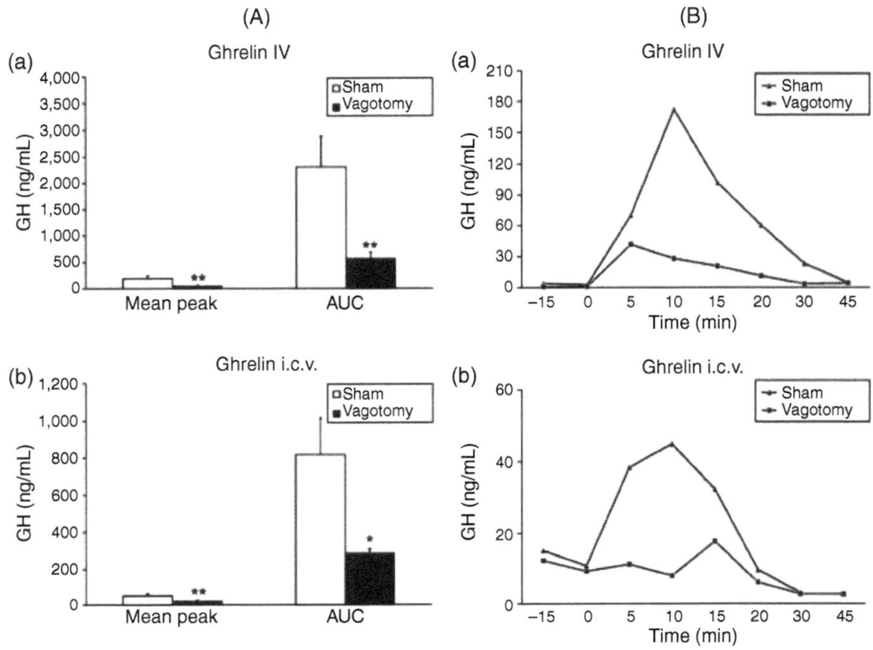

Figure 1 (A) Mean peak GH concentrations and area under the curve (AUC) (mean ± SE) in control and vagotomized rats receiving ghrelin i.v. (20 μg/kg) (a) or i.c.v. (5 μg/rat) (b) ($n = 10$). (B) Representative secretor profile of a control rat treated with ghrelin i.v. (20 μg/kg) (a) or i.c.v. (5 μg/rat), and a vagotomized rat treated with ghrelin i.v. (20 μg/kg) (b) or i.c.v. (5 μg/rat), $^{*}p < 0.05$ and $^{**}p < 0.01$.[51]

together, these results show that the vagus nerve is crucial in GH regulation and is essential for the full GH-releasing effect of ghrelin.

Data from humans showed that ghrelin dose-dependent mediates GH secretion.[52] However, controversy has increased with respect to the relationship between GH and ghrelin circulating levels in humans, showing some a direct relationship between circulating levels of GH and ghrelin,[53] while other studies have not found any relationship between them.[54] The origin of this controversy may be that a majority of these studies did not distinguish total ghrelin from acyl ghrelin. Some authors support the fact that the relationship between ghrelin and GH is only found under certain conditions of nutritional status in fasting states. Accordingly, a relationship between GH and ghrelin pulses was described in the fasting state at night.[53] Detailed study in humans found a correlation between GH pulses and the concentration of acyl ghrelin during fed conditions (Table 1).[55]

Table 1 Evidences Supporting a Physiological Role of Ghrelin on GH Secretion.

Physiological Role	References
Ghrelin induces GH secretion *in vivo* in a more potent fashion than GHRH	[44]
Ghrelin induces GH secretion in humans	[52]
Ghrelin induces GH secretion *in vitro*	[40]
Pituitary stalk lesions decrease GHS actions in rats	[37]
Humans with traumatic brain injuries present decreased GH responses to GHS	[43]
Ghrelin inhibits stimulated somatostatin release	[46]
An intact vagus nerve is required for ghrelin to induce its maximum effect on GH secretion. Vagal blockade decreases GH secretion induced by ghrelin.	[50,51]
GOAT $^{-/-}$ mice under caloric restriction are hypoglycemic due to decreased GH secretion	[70]
Obesity is associated with low ghrelin levels and low GH secretion	[76,77]
Anorexia is associated with high ghrelin levels and high GH secretion	[81]

4. DOES GHRELIN PARTICIPATE IN THE PHYSIOLOGICAL REGULATION OF GH?

Building on the initial findings of the main function of ghrelin as a GH regulator in the splanchnic area, with the development of knockout models for ghrelin, GHS or the physiological effect of ghrelin on GH secretion, has been further questioned. In fact, knockout of ghrelin or the ghrelin receptor showed only a modest decrease in appetite with normal body growth and no differences in GH levels, with respect to wild type, in spite of the lack of ghrelin production.[56–58] In addition, the ghrelin-knockout mice failed to show alterations in feeding and fasting leptin and insulin levels. The lack of ghrelin did not protect against high-fat diet-induced obesity.[56] All together, these data led to the conclusion that ghrelin is not required for physiological functions such as growth, appetite, fertility, bone density, and fat deposition. Accordingly, it was proposed that the antagonist for ghrelin is not very useful in treating obesity and related disorders.

5. LINK BETWEEN NUTRITIONAL STATUS, GHRELIN PRODUCTION, AND GH SECRETION

5.1 Effects of Ghrelin and GH on Glucose Metabolism

Several studies aimed to elucidate the potential role of ghrelin in regulating glucose homeostasis.[59–61] An inverse relationship was described between ghrelin and insulin levels in the blood, as it shown by the fact that insulin infusion decreased ghrelin levels in the plasma, which was recovered down to basal levels after the discontinuation of the infusion.[62] This inverse relationship suggests that insulin is a physiological and dynamic modulator of the plasma ghrelin.

In human studies, it was initially suggested that ghrelin decreases insulin secretion.[63] However, subsequent studies showed that depending on the conditions and dose, ghrelin can increase or decrease insulin levels.[64] It was also reported that ghrelin modulates insulin sensitivity depending on the acyl ghrelin/ghrelin ratio.[65] The relationship between ghrelin and insulin has not been completely clarified yet, and controversial data exist in the literature.[66–69]

It was proposed that ghrelin might act as a glucose sensor, based on the fact that under fasting conditions, the levels of ghrelin, and GHSR in the hypothalamus are elevated. Accordingly, it was suggested that increased acyl ghrelin promotes insulin desensitization,[65] which together with the fact that acyl ghrelin is the active form inducing GH secretion, suggests that the regulatory effect of ghrelin on GH production possibly regulates glucose homeostasis. In fact, GOAT$^{-/-}$ mice were shown to have improved glucose-induced insulin secretion and glucose tolerance.[70] Moreover, GOAT$^{-/-}$ mice under caloric restriction are hypoglycemic due to a decreased GH secretion.

In light of the previously presented data, it would be plausible to conclude that the splanchnic regulation of GH levels would be the part of a complex mechanism, encompassing nutritional status with the regulation of ghrelin and GH, in order to orchestrate glucose homeostasis under altered nutritional status.

In 2008, two different groups identified the enzyme responsible for ghrelin octanoylation, that is GOAT.[71,72] Focused on the fact that ghrelin octanoylation seems to be crucial for its GH-releasing activity, it may be possible that GH-releasing activity would be impaired in GOAT$^{-/-}$ mice. However, the first work to include a model of GOAT$^{-/-}$ mice only found a minor reduction in weight, when the animals were fed a diet rich in octanoylate, showing that GOAT plays a role in mediating the effect of lipid intake on adiposity.[24] In 2011, Zhao *et al.* reported the role of GOAT under caloric restriction by the generation of a GOAT$^{-/-}$ model subjected to 60% caloric restriction for 7 days. Under fed conditions, the elimination of GOAT did not affect growth, but after caloric restriction, a 30% decrease in body weight and 75% decrease in the body fat were observed after 4 days. Moreover, the levels of glucose in fasting states were also reduced in both GOAT$^{-/-}$ and wild-type mice. A great difference was that in wild-type mice, the glucose levels got stabilized, but the KO model animals did not recover from normal glucose levels and ultimately died. Interestingly, it was also observed that the difference between both the models was their GH level after caloric restriction. The GH levels progressively decreased in both wild type and KO models, but in the wild type mice, GH recovered to normal levels after an initial drop. In KO models, this recovery was very low in comparison to that of the wild type, and was not sufficient to increase the glucose levels. These findings were of great relevance in explaining the physiological role of acyl ghrelin on GH regulation. It is widely known that one of the main physiological functions of GH is to maintain glucose levels under negative energy balance conditions by increasing lipolysis in adipose tissue with the release of fatty acids, that spare glucose utilization in the muscle.[73] All together, these data show that a major role of ghrelin in severe caloric-restriction states is to facilitate GH secretion, with the aim of preserving blood glucose and preventing death.[70] Moreover, it was shown that GH maintains blood glucose levels in calorie-restricted mice partly, but not totally, by activating hepatic autophagy because when the fat deposits are completely depleted, the increase in autophagy is not sufficient to maintain blood glucose levels.[74]

In contrast to these data, a subsequent study by Yi *et al.* found that the endogenous goat–ghrelin–GHSR system is not essential for the maintenance of euglycemia during prolonged calorie restriction.[75] However, the differential methodological aspects between Zhao's and Yi's works should be taken into account. For example, Yi used tail vein as the source of the sample to measure glucose levels, while Zhao used retro-orbital bleeding. In addition, composition of the diet was different in terms of carbohydrates. Yi used a diet with a higher carbohydrate composition, which may have caused the prevention of hypoglycemia found in the calorie-restricted mice. Lastly, Zhao's animals were considerably younger than Yi's.

5.2 GH—Ghrelin Status and Obesity

Until now, all the transgenic models produced to disrupt GH signaling yielded dwarf and obese animals. The low GH levels in obese individuals are associated with a worse cardiovascular prognosis. However, low GH levels in obesity are considered to be a functional defect, and although the exact mechanism behind this alteration in obese patients is not well known, the deregulation of GHRH, SS, and ghrelin pathways might be involved. In addition, impairment of the levels of several obesity-related factors, such as free fatty acids and hyperinsulinemia, seems to be directly implied. Circulating ghrelin levels were shown to be decreased in obesity.[76] It was suggested that the low ghrelin levels described in obesity, could mediate the decreased GH secretion of the pathological state.[77] This was based on several studies in humans showing that obese patients have a better GH response to ghrelin or GHRH + ghrelin than those treated only with GHRH.[78] In addition, it was proposed that visceral fat mass is a strong predictor of basal-ghrelin concentration and 24 h GH release.[79,80] In light of these findings, now the question is to determine whether treatment with a ghrelin analogue can recover the decreased GH levels in obesity, which is currently assumed to be improbable due to the adipogenic effect of ghrelin.

5.3 Ghrelin/GH-Induced Anorexia Nervosa

Patients with anorexia nervosa (AN) are characterized by high GH levels and hepatic resistance to GH, which is reflected at very low levels of the circulating IGF-1. It is not known whether the high GH secretion from the pituitary, induced by feedback in the somatotropic axis, was produced by low IGF-1 levels. A direct relationship between ghrelin signaling in AN and altered GH/IGF-1 status was proposed.[81] In spite of the primary role of ghrelin in increasing appetite and food intake, its circulating levels are elevated in AN, which suggests the existence of a state of ghrelin resistance in these patients. However, the published data from studies, which tested the potential role of ghrelin in increasing GH secretion in AN patients, are controversial. Some studies did not find a correlation between GH and circulating ghrelin levels.[82] In patients with AN, the GH response to GHRH was found to be increased, suggesting that this effect is mediated by the increased circulating levels of ghrelin, described in this pathology. In support of this relationship, a study showed that an adolescent woman with AN presents circulating ghrelin levels that predicted a GH burst.[83]

6. A NOVEL PERIPHERAL MECHANISM MEDIATING THE REGULATION OF GH SECRETION BY GHRELIN; ROLE OF GASTRIC CANNABINOID SYSTEM

GH and somatotropic axis are directly involved in the regulation of metabolism. In the past years, the cannabinoid system has been described as a novel system involved in both regulation of energy homeostasis and the neuroendocrine axis.[84] In the past, several studies were focused on the neuroendocrine effects of the main psychoactive compound of marijuana, the Δ9-tetrahydrocannabinol (Δ9-THC).[84] The pharmacological regulation of the cannabinoid receptor 1 (CB1) has been studied in depth, due to its actions on energy homeostasis and neuroendocrine axis. The main connection between the cannabinoid system and somatotropic axis was shown by several studies, which focused on the effect of Δ9-THC and synthetic cannabinoids on GH secretion, showing that these compounds inhibit GH secretion in rodents after acute administration.[85–87] In humans, it was shown that the consumption of marijuana decreases the response of GH to insulin-induced hypoglycemia.[88] Insufficient data exist to identify whether the regulatory action of cannabinoids on GH secretion is produced at the central level or if the cannabinoids in the peripheral organs are also involved in this process. Several studies have led to speculation that the activation of hypothalamic SS production is responsible for the inhibitory action of Δ9-THC on GH secretion.[89]

A cross-talk interaction at the peripheral level was recently described, most concretely in the gastrointestinal tract, between the cannabinoid system and ghrelin. CB1 expression was described in the gastric fundus and was localized in the same cells involved in ghrelin production at this level. This finding highlighted the possibility of a functional interaction between these systems in the regulation of food intake and GH production. By using a gastric explant model, it was shown that the pharmacological blockade of gastric CB1 reverts the increased production of ghrelin, which is characteristic of the fasting state and reduces circulating ghrelin levels. In addition, the work found that this interaction disappears when the blockade of peripheral CB1 was performed in animals under surgical vagotomy, indicating that communication via vagus nerve is crucial for the interaction between ghrelin and cannabinoids, in the regulation of energy homeostasis. After the gastric interaction between the cannabinoid system and ghrelin was shown to be involved in food intake and the regulation of energy balance, it was shown that the interaction between these two systems might constitute a novel mechanism of GH regulation by the splanchnic area. After confirming this theory, it was found that cannabinoid signals are constitutively active in the pathways that control GH secretion, as shown by the fact that the pharmacological peripheral blockade of the CB1 receptor induces a decrease in pulsatile GH secretion and a decreased GH response to exogenous ghrelin administration. Moreover, the inactivation of peripheral CB1 induced a decrease in GHRH mRNA expression at the hypothalamic level. In addition, the GH response to ghrelin administration in animals with peripheral CB1 blockade by intraperitoneal rimonabant treatment was significantly reduced, supposedly as a consequence of the downregulation of GHRH pathway after cannabinoid modulation. In support of the regulatory effect of the peripheral cannabinoid system on GH, it was shown that the central administration of rimonabant, the antagonist of CB1 receptor, is unable to modify GHRH or SS mRNA at the hypothalamic level or response of GH to ghrelin (Fig. 2).

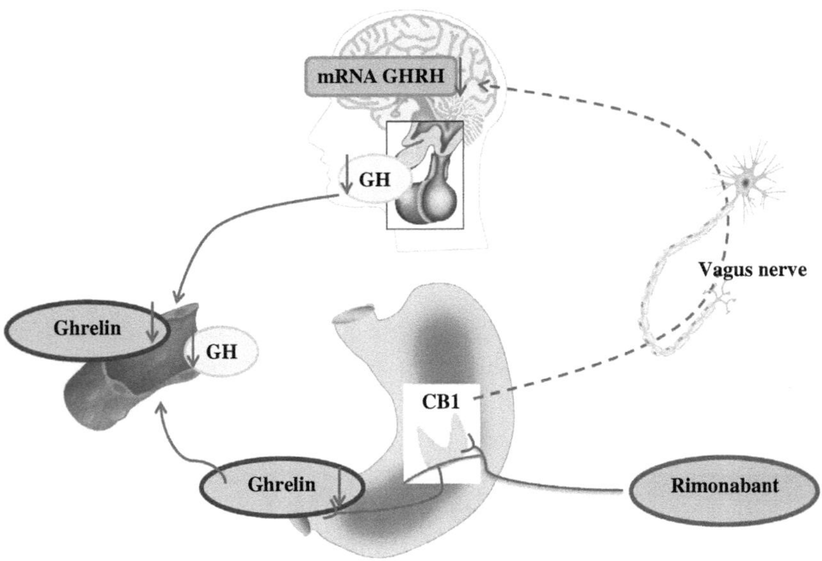

Figure 2 Functional interactions between cannabinoid and ghrelin, to regulate GH secretion. The pharmacological blockade of gastric CB1 by peripheral rimonabant injection decrease gastric ghrelin secretion, circulating ghrelin levels, and the pulsatile GH secretion. Moreover, the inactivation of peripheral CB1 by rimonabant, decreases the GH response to ghrelin exogenously administered, and also decreases GHRH mRNA expression at a hypothalamic level. The vagus nerve communication is crucial for the interaction between ghrelin and cannabinoid system, in the regulation of GH.

A great relevance of this new mechanism is highlighted by the fact that the somatotropic regulation of cannabinoids should be considered, when designing cannabinoid-derived drugs to treat obesity. The modulation of GH by cannabinoids would probably generate some psychiatric side effects, as have been recorded in obese patients treated with these drugs.[90] The central actions of GH on the mood and well-being, have been described and apparently, GH gets reduced in depression.[91]

In this context, a strong link has been proposed among nutritional status, ghrelin production, and GH secretion. Data in humans supporting this complex relationship are based on several facts: first, ghrelin and GH plasma levels are increase greatly in AN patients. Second, anorexia is characterized by severe undernutrition. Third, a potential action of acyl ghrelin has been proposed in mice, maintaining elevated plasma GH levels during periods of severe calorie restriction and thereby assuring sufficient blood glucose to permit survival. All together, these findings yielded the following question: does ghrelin increase GH in anorexia patients to maintain glucose levels in the appropriate range for survival? Further studies should be conducted to answer this question.

Thus, ghrelin participates in GH regulation, especially under extreme conditions. A complex system at the gastric level monitors the nutritional status of the organism and modulates ghrelin production. This information is conveyed to the brain by the vagus nerve, and the brain responds through the regulation of a somatotropic axis at the hypothalamic and pituitary levels, in order to restore the energy balance. From a physiological standpoint, a novel mechanism of GH regulation by ghrelin exists, implicating the peripheral modulation of the CB.

REFERENCES

1. Kopchick JJ, Parkinson C, Stevens EC, Trainer PJ. Growth hormone receptor antagonists: discovery, development, and use in patients with acromegaly. *Endocr Rev.* 2002;23(5):623–646.
2. Daughaday WH. A personal history of the origin of the somatomedin hypothesis and recent challenges to its validity. *Perspect Biol Med.* 1989;32(2):194–211.
3. Baumann G. Metabolism of growth hormone (GH) and different molecular forms of GH in biological fluids. *Horm Res.* 1991;36(Suppl 1):5–10.
4. Lanning NJ, Carter-Su C. Recent advances in growth hormone signaling. *Rev Endocr Metab Disorders.* 2006;7(4):225–235.
5. Holloway MG, Cui Y, Laz EV, Hosui A, Hennighausen L, Waxman DJ. Loss of sexually dimorphic liver gene expression upon hepatocyte-specific deletion of Stat5a-Stat5b locus. *Endocrinology.* 2007;148(5):1977–1986.
6. Savastano S, Di Somma C, Barrea L, Colao A. The complex relationship between obesity and the somatropic axis: the long and winding road. *Growth Horm IGF Res.* 2014;24(6):221–226.

7. Tsukazaki T, Matsumoto T, Enomoto H, et al. Growth hormone directly and indirectly stimulates articular chondrocyte cell growth. *Osteoarthr Cartil.* 1994;2(4):259–267.
8. Colao A, Merola B, Ferone D, Lombardi G. Acromegaly. *J ClinEndocrinol Metab.* 1997;82 (9):2777–2781.
9. Ho KK, O'Sullivan AJ, Hoffman DM. Metabolic actions of growth hormone in man. *EndocrJ.* 1996;43(Suppl):S57–S63.
10. Russell-Jones DL, Weissberger AJ. The role of growth hormone in the regulation of body composition in the adult. *Growth Regul.* 1996;6(4):247–252.
11. Nam SY, Lobie PE. The mechanism of effect of growth hormone on preadipocyte and adipocyte function. *Obesity Rev.* 2000;1(2):73–86.
12. Vijayakumar A, Yakar S, Leroith D. The intricate role of growth hormone in metabolism. *Frontiers Endocrinol.* 2011;2:32.
13. Richelsen B. Action of growth hormone in adipose tissue. *Horm Res.* 1997;48(Suppl 5): 105–110.
14. Richelsen B, Pedersen SB, Kristensen K, et al. Regulation of lipoprotein lipase and hormone-sensitive lipase activity and gene expression in adipose and muscle tissue by growth hormone treatment during weight loss in obese patients. *Metab ClinExp.* 2000;49 (7):906–911.
15. Zhao JT, Cowley MJ, Lee P, Birzniece V, Kaplan W, Ho KK. Identification of novel GH-regulated pathway of lipid metabolism in adipose tissue: a gene expression study in hypopituitary men. *J Clin Endocrinol Metab.* 2011;96(7):E1188–E1196.
16. Li Y, Knapp JR, Kopchick JJ. Enlargement of interscapular brown adipose tissue in growth hormone antagonist transgenic and in growth hormone receptor gene-disrupted dwarf mice. *Exp Biol Med (Maywood).* 2003;228(2):207–215.
17. Wu Y, Liu C, Sun H, et al. Growth hormone receptor regulates beta cell hyperplasia and glucose-stimulated insulin secretion in obese mice. *J Clin Invest.* 2011;121(6):2422–2426.
18. Ohlsson C, Mohan S, Sjogren K, et al. The role of liver-derived insulin-like growth factor-I. *Endocr Rev.* 2009;30(5):494–535.
19. Romero CJ, Ng Y, Luque RM, et al. Targeted deletion of somatotroph insulin-like growth factor-I signaling in a cell-specific knockout mouse model. *Mol Endocrinol.* 2010;24(5):1077–1089.
20. Bowers CY. Unnatural growth hormone-releasing peptide begets natural ghrelin. *J Clin Endocrinol Metab.* 2001;86(4):1464–1469.
21. Kojima M, Hosoda H, Date Y, Nakazato M, Matsuo H, Kangawa K. Ghrelin is a growth-hormone-releasing acylated peptide from stomach. *Nature.* 1999;402(6762):656–660.
22. Ariyasu H, Takaya K, Tagami T, et al. Stomach is a major source of circulating ghrelin, and feeding state determines plasma ghrelin-like immunoreactivity levels in humans. *J Clin Endocrinol Metab.* 2001;86(10):4753–4758.
23. Dornonville de la Cour C, Bjorkqvist M, Sandvik AK, et al. A-like cells in the rat stomach contain ghrelin and do not operate under gastrin control. *Regul Peptides.* 2001;99(2-3):141–150.
24. Kirchner H, Gutierrez JA, Solenberg PJ, et al. GOAT links dietary lipids with the endocrine control of energy balance. *Nat Med.* 2009;15(7):741–745.
25. Tschop M, Smiley DL, Heiman ML. Ghrelin induces adiposity in rodents. *Nature.* 2000;407(6806):908–913.
26. Al-Massadi O, Crujeiras AB, Gonzalez RC, et al. Age, sex, and lactating status regulate ghrelin secretion and GOAT mRNA levels from isolated rat stomach. *Am J Physiol Endocrinol Metab.* 2010;299(3):E341–E350.
27. Al Massadi O, Tschop MH, Tong J. Ghrelin acylation and metabolic control. *Peptides.* 2011;32(11):2301–2308.
28. Cummings DE, Weigle DS, Frayo RS, et al. Plasma ghrelin levels after diet-induced weight loss or gastric bypass surgery. *New Engl J Med.* 2002;346(21):1623–1630.

29. Seoane LM, Al-Massadi O, Caminos JE, Tovar SA, Dieguez C, Casanueva FF. Sensory stimuli directly acting at the central nervous system regulate gastric ghrelin secretion. An *ex vivo* organ culture study. *Endocrinology.* 2007;148(8):3998–4006.
30. Sawchenko PE. Central connections of the sensory and motor nuclei of the vagus nerve. *J Autonom Nervous Sys.* 1983;9(1):13–26.
31. Seoane LM, Al-Massadi O, Barreiro F, Dieguez C, Casanueva FF. Growth hormone and somatostatin directly inhibit gastric ghrelin secretion. An *in vitro* organ culture system. *J Endocrinol Invest.* 2007;30(9):RC22–RC25.
32. Di Vito L, Broglio F, Benso A, et al. The GH-releasing effect of ghrelin, a natural GH secretagogue, is only blunted by the infusion of exogenous somatostatin in humans. *Clin Endocrinol.* 2002;56(5):643–648.
33. Patel YC. Somatostatin and its receptor family. *Front Neuroendocrinol.* 1999;20(3): 157–198.
34. Broglio F, Koetsveld Pv P, Benso A, et al. Ghrelin secretion is inhibited by either somatostatin or cortistatin in humans. *J Clin Endocrinol Metab.* 2002;87(10):4829–4832.
35. Lopez-Siguero JP, Lopez-Canti LF, Espino R, et al. Effect of recombinant growth hormone on leptin, adiponectin, resistin, interleukin-6, tumor necrosis factor-alpha and ghrelin levels in growth hormone-deficient children. *J Endocrinol Investig.* 2011;34 (4):300–306.
36. Nagano M, Chastre E, Choquet A, Bara J, Gespach C, Kelly PA. Expression of prolactin and growth hormone receptor genes and their isoforms in the gastrointestinal tract. *Am J Physiol.* 1995;268(3 Pt 1):G431–G442.
37. van der Lely AJ, Tschop M, Heiman ML, Ghigo E. Biological, physiological, pathophysiological, and pharmacological aspects of ghrelin. *Endocr Rev.* 2004;25(3): 426–457.
38. Klok MD, Jakobsdottir S, Drent ML. The role of leptin and ghrelin in the regulation of food intake and body weight in humans: a review. *Obes Rev.* 2007;8(1):21–34.
39. Ghigo E, Broglio F, Arvat E, Maccario M, Papotti M, Muccioli G. Ghrelin: more than a natural GH secretagogue and/or an orexigenic factor. *Clin Endocrinol.* 2005;62(1):1–17.
40. Popovic V, Miljic D, Micic D, et al. Ghrelin main action on the regulation of growth hormone release is exerted at hypothalamic level. *J Clin Endocrinol Metab.* 2003;88 (7):3450–3453.
41. Bowers CY, Sartor AO, Reynolds GA, Badger TM. On the actions of the growth hormone-releasing hexapeptide, GHRP. *Endocrinology.* 1991;128(4):2027–2035.
42. Clark RG, Carlsson LM, Trojnar J, Robinson IC. The effects of a growth hormone-releasing peptide and growth hormone-releasing factor in conscious and anaesthetized rats. *J Neuroendocrinol.* 1989;1(4):249–255.
43. Pavlovic D, Pekic S, Stojanovic M, et al. Chronic cognitive sequelae after traumatic brain injury are not related to growth hormone deficiency in adults. *Eur J Neurol.* 2010;17 (5):696–702.
44. Seoane LM, Tovar S, Baldelli R, et al. Ghrelin elicits a marked stimulatory effect on GH secretion in freely-moving rats. *Eur J Endocrinol.* 2000;143(5):R7–R9.
45. Kamegai J, Wakabayashi I, Kineman RD, Frohman LA. Growth hormone-releasing hormone receptor (GHRH-R) and growth hormone secretagogue receptor (GHS-R) mRNA levels during postnatal development in male and female rats. *J Neuroendocrinol.* 1999;11(4):299–306.
46. Tolle V, Zizzari P, Tomasetto C, Rio MC, Epelbaum J, Bluet-Pajot MT. *In vivo* and *in vitro* effects of ghrelin/motilin-related peptide on growth hormone secretion in the rat. *Neuroendocrinology.* 2001;73(1):54–61.
47. Tannenbaum GS, Epelbaum J, Bowers CY. Interrelationship between the novel peptide ghrelin and somatostatin/growth hormone-releasing hormone in regulation of pulsatile growth hormone secretion. *Endocrinology.* 2003;144(3):967–974.

48. Seoane LM, Lopez M, Tovar S, Casanueva FF, Senaris R, Dieguez C. Agouti-related peptide, neuropeptide Y, and somatostatin-producing neurons are targets for ghrelin actions in the rat hypothalamus. *Endocrinology.* 2003;144(2):544–551.
49. Arvat E, Maccario M, Di Vito L, et al. Endocrine activities of ghrelin, a natural growth hormone secretagogue (GHS), in humans: comparison and interactions with hexarelin, a nonnatural peptidyl GHS, and GH-releasing hormone. *J Clin Endocrinol Metab.* 2001;86 (3):1169–1174.
50. Date Y, Murakami N, Toshinai K, et al. The role of the gastric afferent vagal nerve in ghrelin-induced feeding and growth hormone secretion in rats. *Gastroenterology.* 2002;123(4):1120–1128.
51. Al-Massadi O, Trujillo ML, Senaris R, et al. The vagus nerve as a regulator of growth hormone secretion. *Regul Pept.* 2011;166(1-3):3–8.
52. Peino R, Baldelli R, Rodriguez-Garcia J, et al. Ghrelin-induced growth hormone secretion in humans. *Eur J Endocrinol.* 2000;143(6):R11–R14.
53. Koutkia P, Canavan B, Breu J, Johnson ML, Grinspoon SK. Nocturnal ghrelin pulsatility and response to growth hormone secretagogues in healthy men. *Am J Physiol Endocrinol Metab.* 2004;287(3):E506–E512.
54. Avram AM, Jaffe CA, Symons KV, Barkan AL. Endogenous circulating ghrelin does not mediate growth hormone rhythmicity or response to fasting. *J Clin Endocrinol Metab.* 2005;90(5):2982–2987.
55. Nass R, Farhy LS, Liu J, et al. Evidence for acyl-ghrelin modulation of growth hormone release in the fed state. *J Clin Endocrinol Metab.* 2008;93(5):1988–1994.
56. Sun Y, Ahmed S, Smith RG. Deletion of ghrelin impairs neither growth nor appetite. *Mol Cell Biol.* 2003;23(22):7973–7981.
57. Wortley KE, Anderson KD, Garcia K, et al. Genetic deletion of ghrelin does not decrease food intake but influences metabolic fuel preference. *Proc Natl Acad Sci USA.* 2004;101 (21):8227–8232.
58. Zigman JM, Nakano Y, Coppari R, et al. Mice lacking ghrelin receptors resist the development of diet-induced obesity. *J Clin Invest.* 2005;115(12):3564–3572.
59. Delhanty PJ, van der Lely AJ. Ghrelin and glucose homeostasis. *Peptides.* 2011;32 (11):2309–2318.
60. Briggs DI, Andrews ZB. A recent update on the role of ghrelin in glucose homeostasis. *Curr Diab Rev.* 2011;7(3):201–207.
61. Verhulst PJ, Janssen S, Tack J, Depoortere I. Role of the AMP-activated protein kinase (AMPK) signaling pathway in the orexigenic effects of endogenous ghrelin. *Regul Peptides.* 2012;173(1-3):27–35.
62. Saad RJ, Danadian K, Lewy V, Arslanian SA. Insulin resistance of puberty in African-American children: lack of a compensatory increase in insulin secretion. *Pediatr Diab.* 2002;3(1):4–9.
63. Broglio F, Arvat E, Benso A, et al. Ghrelin, a natural GH secretagogue produced by the stomach, induces hyperglycemia and reduces insulin secretion in humans. *J Clin Endocrinol Metab.* 2001;86(10):5083–5086.
64. Salehi A, Dornonville de la Cour C, Hakanson R, Lundquist I. Effects of ghrelin on insulin and glucagon secretion: a study of isolated pancreatic islets and intact mice. *Regul Pept.* 2004;118(3):143–150.
65. Barazzoni R, Zanetti M, Ferreira C, et al. Relationships between desacylated and acylated ghrelin and insulin sensitivity in the metabolic syndrome. *J Clin Endocrinol Metab.* 2007;92(10):3935–3940.
66. Flanagan DE, Evans ML, Monsod TP, et al. The influence of insulin on circulating ghrelin. *Am J Physiol Endocrinol Metab.* 2003;284(2):E313–E316.
67. Lucidi P, Murdolo G, Di Loreto C, et al. Ghrelin is not necessary for adequate hormonal counterregulation of insulin-induced hypoglycemia. *Diabetes.* 2002;51(10):2911–2914.

68. Mohlig M, Spranger J, Otto B, Ristow M, Tschop M, Pfeiffer AF. Euglycemic hyperinsulinemia, but not lipid infusion, decreases circulating ghrelin levels in humans. *J Endocrinol Investig.* 2002;25(11):RC36–RC38.
69. Caixas A, Bashore C, Nash W, Pi-Sunyer F, Laferrere B. Insulin, unlike food intake, does not suppress ghrelin in human subjects. *J Clinic Endocrinol Metab.* 2002;87(4):1902.
70. Zhao TJ, Liang G, Li RL, et al. Ghrelin O-acyltransferase (GOAT) is essential for growth hormone-mediated survival of calorie-restricted mice. *Proc Natl Acad Sci USA.* 2010;107 (16):7467–7472.
71. Gutierrez JA, Solenberg PJ, Perkins DR, et al. Ghrelin octanoylation mediated by an orphan lipid transferase. *Proc Natl Acad Sci USAmerica.* 2008;105(17):6320–6325.
72. Yang J, Brown MS, Liang G, Grishin NV, Goldstein JL. Identification of the acyltransferase that octanoylates ghrelin, an appetite-stimulating peptide hormone. *Cell.* 2008;132 (3):387–396.
73. Moller N, Jorgensen JO. Effects of growth hormone on glucose, lipid, and protein metabolism in human subjects. *Endocr Rev.* 2009;30(2):152–177.
74. Zhang Y, Fang F, Goldstein JL, Brown MS, Zhao TJ. Reduced autophagy in livers of fasted, fat-depleted, ghrelin-deficient mice: reversal by growth hormone. *Proc Natl Acad Sci USA.* 2015;112(4):1226–1231.
75. Yi CX, Heppner KM, Kirchner H, et al. The GOAT-ghrelin system is not essential for hypoglycemia prevention during prolonged calorie restriction. *PloS ONE.* 2012;7(2):e32100.
76. Tschop M, Weyer C, Tataranni PA, Devanarayan V, Ravussin E, Heiman ML. Circulating ghrelin levels are decreased in human obesity. *Diabetes.* 2001;50(4):707–709.
77. Nass R, Gaylinn BD, Thorner MO. The role of ghrelin in GH secretion and GH disorders. *Mol Cell Endocrinol.* 2011;340(1):10–14.
78. Alvarez-Castro P, Isidro ML, Garcia-Buela J, et al. Marked GH secretion after ghrelin alone or combined with GH-releasing hormone (GHRH) in obese patients. *Clin Endocrinol.* 2004;61(2):250–255.
79. Katsuki A, Urakawa H, Gabazza EC, et al. Circulating levels of active ghrelin is associated with abdominal adiposity, hyperinsulinemia and insulin resistance in patients with type 2 diabetes mellitus. *Eur J Endocrinol.* 2004;151(5):573–577.
80. Clasey JL, Weltman A, Patrie J, et al. Abdominal visceral fat and fasting insulin are important predictors of 24-hour GH release independent of age, gender, and other physiological factors. *J Clin Endocrinol Metab.* 2001;86(8):3845–3852.
81. Gianotti L, Lanfranco F, Ramunni J, Destefanis S, Ghigo E, Arvat E. GH/IGF-I axis in anorexia nervosa. *EWD.* 2002;7(2):94–105.
82. Tolle V, Bassant MH, Zizzari P, et al. Ultradian rhythmicity of ghrelin secretion in relation with GH, feeding behavior, and sleep-wake patterns in rats. *Endocrinology.* 2002;143 (4):1353–1361.
83. Misra M, Miller KK, Kuo K, et al. Secretory dynamics of ghrelin in adolescent girls with anorexia nervosa and healthy adolescents. *Am J Physiol Endocrinol Metab.* 2005;289(2): E347–E356.
84. Pagotto U, Marsicano G, Cota D, Lutz B, Pasquali R. The emerging role of the endocannabinoid system in endocrine regulation and energy balance. *Endocr Rev.* 2006;27(1):73–100.
85. Dalterio SL, Michael SD, Macmillan BT, Bartke A. Differential effects of cannabinoid exposure and stress on plasma prolactin, growth hormone and corticosterone levels in male mice. *Life Sci.* 1981;28(7):761–766.
86. Kokka N, Garcia JF. Effects of delta 9-THC on growth hormone and ACTH secretion in rats. *Life Sci.* 1974;15(2):329–338.
87. Martin-Calderon JL, Munoz RM, Villanua MA, et al. Characterization of the acute endocrine actions of (-)-11-hydroxy-delta8-tetrahydrocannabinol-dimethylheptyl (HU-210), a potent synthetic cannabinoid in rats. *Eur J Pharmacol.* 1998;344(1):77–86.

88. Benowitz NL, Jones RT, Lerner CB. Depression of growth hormone and cortisol response to insulin-induced hypoglycemia after prolonged oral delta-9-tetrahydrocannabinol administration in man. *J Clin Endocrinol Metab.* 1976;42(5):938–941.
89. Rettori V, Aguila MC, Gimeno MF, Franchi AM, McCann SM. *Invitro* effect of delta 9-tetrahydrocannabinol to stimulate somatostatin release and block that of luteinizing hormone-releasing hormone by suppression of the release of prostaglandin E2. *Proc Natl Acad Sci USA.* 1990;87(24):10063–10066.
90. Christensen R, Kristensen PK, Bartels EM, Bliddal H, Astrup A. Efficacy and safety of the weight-loss drug rimonabant: a meta-analysis of randomised trials. *Lancet.* 2007;370 (9600):1706–1713.
91. Schneider HJ, Pagotto U, Stalla GK. Central effects of the somatotropic system. *Eur J Endocrinol.* 2003;149(5):377–392.

SECTION II

Acromegaly

CHAPTER FOUR

The Modern Criteria for Medical Management of Acromegaly

Stefano Frara, Filippo Maffezzoni, Gherardo Mazziotti, Andrea Giustina[1]
Endocrinology, University of Brescia, Brescia, Italy
[1] Corresponding author: e-mail address: a.giustina@libero.it

Contents

Abstract

Acromegaly is an insidious disorder characterized by excess secretion of growth hormone (GH) and elevated circulating levels of insulin-like growth factor-I (IGF-I), generally caused by a pituitary adenoma. It is a rare disease associated with an average 10-year reduction in life expectancy due to metabolic, cardiovascular, and cerebrovascular comorbidities and reduced quality of life caused by paresthesias, fatigue, osteoarthralgia, or bone fractures. In 2000, Cortina Consensus Conference established general criteria for diagnosis and biochemical control of acromegaly, which have been revised in recent years, adapting them to emerging clinical evidences as well as the evolving assay techniques. Authors have proposed a binary definition of cure for acromegaly, where both GH and IGF-I are important determinants: the former is more linked to the presence of residual adenomatous tissue, while the latter to the peripheral activity of the disease. Control of tumor growth and complications is also an essential goal of treatment. Surgical, medical, and radiotherapy approaches are all valid alternatives. The surgical option is, however, unsuccessful in about 50% of patients. Somatostatin analogs (SRLs), octreotide LAR, and lanreotide ATG can inhibit cell growth, besides their beneficial effects

Progress in Molecular Biology and Translational Science, Volume 138
ISSN 1877-1173
http://dx.doi.org/10.1016/bs.pmbts.2015.10.015

on GH hypersecretion and on most comorbidities. Pasireotide is a new multireceptor-targeted SRL with reported superior biochemical efficacy to octreotide, due to higher affinity for SSTR-5, but potentially causing detrimental effects on glucose homeostasis. Pegvisomant could be a valid choice in all patients resistant to SRLs. It is a competitive GH antagonist, which efficaciously blocks IGF-I production, inhibiting the dimerization of GH receptor. Normal IGF-I levels represent, therefore, its only relevant efficacy endpoint, while only few cases of tumor growth on pegvisomant have been reported, so far.

1. INTRODUCTION

Acromegaly is an insidious disorder characterized by excess secretion of growth hormone (GH) generally caused by a pituitary macroadenoma (~70% of cases) resulting in elevations of circulating levels of GH and insulin-like growth factor (IGF-I).[1] Although it is a rare disease, acromegaly is associated with an average 10-year reduction in life expectancy and a double standardized mortality rate as compared to the general population, due to metabolic, cardiovascular, cerebrovascular, and respiratory comorbidities.[2,3] Management of patients with acromegaly is often problematic, complex, costly, and requires approaches tailored to each individual patient. In fact, a multimodal approach is often required to control acromegaly by suppressing GH hypersecretion, reducing IGF-I levels, and controlling tumor growth, leading to symptom control and minimizing the associated clinical signs and comorbidities.[4,5]

Over the last 15 years, international guidelines based on expert opinions and literature evidence have been proposed in order to provide recommendations for benchmarking management of acromegaly.

Before the year 2000 there was a wide variability in the criteria arbitrarily used for defining biochemical control of acromegaly in different settings. The first historical step to define the biochemical control of acromegaly was the Cortina Consensus Conference, which for the first time established general criteria for universal use,[6] according to the clinical and epidemiologic evidence of a close relationship between serum hormonal values and clinical outcome of acromegaly. Subsequent consensus conferences in more recent years have adapted these criteria to the evolving assay techniques as well as to the emerging clinical evidences.[7,8]

This chapter deals with the therapeutic approach to acromegaly based on the modern criteria for diagnosis and treatment of acromegaly, with specific focus on medical management.

2. THERAPEUTIC END-POINTS

Objectives of acromegaly treatment are normalization of GH hypersecretion and IGF-I values, control of tumor growth, and restoration of morbidity and mortality rate to that of control population.[4,6,7]

2.1 Biochemical Goals

The Cortina Consensus proposed for the first time a binary definition of cure of acromegaly, based on the clinical evidence that both elevated GH and IGF-I were important determinants of outcome of disease in acromegalic patients.[6] From the pathophysiologic point of view, in the patient with acromegaly, GH and IGF-I could give different but complementary information being one parameter (GH) more linked to the presence of residual adenomatous tissue and the other (IGF-I) to the peripheral activity of the hormone secreted by the adenoma. From the clinical point of view, moreover, the normalization of serum IGF-I and the achievement of "safe" GH are both important to reduce the symptoms, control comorbidities, and normalize the mortality risk associated with acromegaly.[3] However, in clinical practice, serum GH and IGF-I levels may give to the clinicians discrepant information in up to 35% of acromegalic patients either after neurosurgery or during somatostatin receptor ligands (SRLs).[9–11] In fact, one parameter may fall in the normal range whereas the other still remains elevated. Frequently, normal GH levels are found in the presence of elevated IGF-I values, but in other cases normal IGF-I associated with high GH values may be observed.[10,11] Several mechanisms may be involved in the determination of this discordance between GH and IGF-I in acromegalic patients and genetic (GH receptor polymorphism) or assay-dependent factors may importantly influence the correct interpretation of biochemical data.[9–12]

Cortina Consensus defined the disease cured, or controlled, when GH nadir during OGTT was below 1 μg/L.[6] In 2010, the guidelines were revised and treatment-specific GH cut-offs were proposed. Revised guidelines proposed 0.4 μg/L as GH cut-off during OGTT together with normalized IGF-I to define cured acromegaly after neurosurgery. Measurement of nadir GH after OGTT was shown to provide reliable information on the biochemical status of disease after neurosurgery but not during medical therapies of acromegaly.[9] Therefore, patients with random GH < 1 μg/L and normal IGF-I were considered controlled after SRLs. Finally, GH cannot be considered a reliable parameter of biochemical control during pegvisomant and patients under this treatment can be considered controlled when IGF-I falls into the normal range for age.[5,7]

Another matter of discussion has been the possible discrepancy in GH and IGF-I values obtained in different laboratories, even when the newest immunoassays are used. From this point of view, the choice of laboratory is crucial for a correct definition of acromegaly status in each patient during the history of disease and in the revision of Cortina criteria the standardization of GH and IGF-I assays has been greatly emphasized.[5,7] Specifically, the revised guidelines recommended the use of the World Health Organization (WHO) international standard (IS 98/574) for measuring the 22-kDa GH isoform[13] and the highly purified international standard (IS 02/254) for the IGF-I assay. Moreover, in the revision of Cortina criteria, the need for each laboratory to have IGF-I normative data drawn from a statistically valid (more than 1000 people) control population stratified by decades was underlined.[7]

However, even when standardization is obtained with the newer methods, discrepancies in measured hormonal levels may be observed. In fact, serum GH and IGF-I values may be influenced by individual factors, such age, sex, body composition, and coexistent systemic diseases which should be taken into account for a correct interpretation of biochemical data in patients with acromegaly.[14]

A concept recently emerged has been the discrimination between partial response and poor response or resistance to SRLs. Partial responders are those patients in whom SRLs induce a clinically significant decrease of GH and/or IGF-I values (i.e., at least 50% of reduction in respect of baseline values) without achieving optimal biochemical targets, whereas nonresponders are patients in whom serum GH and IGF-I are persistently high during SRL treatment.[15] The clinical relevance of such a distinction is related to the possibility of shift from partial to a full response either by increasing the drug doses over the conventional schedules (see Section 2.1) or by replacing the first generation SRLs with the second generation SRL pasireotide (see Section 2.2).

2.2 Tumor Shrinkage and Clinical Goals

Besides the biochemical control of GH hypersecretion, the correct management of acromegaly should include also the evaluation of clinical end-points.

Control of tumor growth is an essential goal of acromegaly treatment and one of the most important advances in this field has been the understanding of the relevant shrinkage effect of SRLs on the GH-secreting adenoma.[16,17] The relevance of this effect has become clear after the widespread use of real long-acting SRLs as the first-line therapy of acromegaly. In fact, long-acting SRLs have been proposed as an alternative tool to neurosurgery in the primary treatment of acromegaly in selected patients (see next section), and in this clinical setting the control of both secretory activity and growth of the tumor are needed for a complete therapeutic success.[18]

GH hypersecretion in acromegaly causes symptoms such as headache, perspiration, paresthesias, fatigue, and osteoarthralgia. Moreover, high GH and IGF-I levels are responsible for the development of chronic complications, such as cardiomyopathy, hypertension, diabetes mellitus, dyslipidemia, sleep apnea, osteoarthropathy, and fragility vertebral fractures (VFs),[19,20] which may lead to an impairment of quality of life and shortened life expectancy:[4] therefore, all patients should be screened for them at diagnosis and during follow-up. In this regard, the use of clinician-reported outcomes may be helpful for therapeutic decision-making.

Dyslipidemia and type 2 diabetes mellitus are often found in patients with acromegaly in close relationship with GH hypersecretion. GH excess causes insulin resistance by an increase in free-fatty acid production with consequent impairment of insulin-stimulated glucose uptake in muscles.[21] Moreover, impairment of pancreatic β-cell function was described in insulin-resistant patients with acromegaly.[22] Diabetes may be, in some patients, the first clinical presentation of acromegaly triggering the diagnostic work-up of GH hypersecretion and should be diagnosed and followed-up based on available criteria for general population.[23,24] The diagnostic criteria for diabetes mellitus in patients with acromegaly are not different from those applied to the general population. In acromegaly, GH excess and insulin resistance concur to determine an "atherogenic dyslipidemia" characterized by high triglycerides and lipoprotein-a (Lp-a), and lower high-density lipoprotein-cholesterol (HDL-C) concentrations.[25,26]

Hypertension is very common in acromegaly (it is found in about half of the patients with active acromegaly).[19,20] Potential pathogenetic mechanisms include sodium-fluid retention, peripheral vasomotor dysfunction, and endothelial disturbances. Guidelines recommend measuring blood pressure at least every 6 months, or during every clinical visit, or after treatment modifications.[20] Twenty-four hour ambulatory blood pressure monitoring (ABPM) could be a very useful instrument for clinicians in order to evaluate a more defined blood pressure profile, and should be assessed annually.[20]

The finding of left-ventricular hypertrophy also in normotensive acromegaly patients is common. Echocardiography has an important role in identifying patients with acromegaly and cardiomyopathy, allowing also to characterize the stage of cardiac disease. In fact, effective acromegaly treatment may induce very rapid improvements of cardiac structural and functional parameters.[27]

Sleep apnea is a common and rather underestimated complication of acromegaly.[28] Every patient should have a careful symptomatic assessment (e.g., by Epworth score), and if necessary laboratory assessment, for sleep apnea at the time of diagnosis and also during treatment of acromegaly, in collaboration with a respiratory physician.[29]

Consistently with the concept that GH excess stimulates bone remodeling, markers of bone formation and resorption are increased in patients with active acromegaly with consequent bone loss and high risk of radiological VFs.[30] The occurrence of VFs in acromegaly correlated with the duration of active disease and serum IGF-I levels, but not with BMD, since they may occur even in patients with normal or low-normal BMD.[30–32] Appropriate and effective treatment of acromegaly improves skeletal health, although the risk of VFs may persist high in some patients with well controlled or cured acromegaly in relationship with pre-existent VFs and untreated hypogonadism.[33] Therefore, guidelines for the diagnosis and follow-up of acromegaly complications recommend to evaluate patients with acromegaly, regardless of BMD values assessed by DXA, with thoracic and lumbar spine X-ray (quantitative morphometry).[20] Morphometric approach may also be applied to images of the spine acquired by DXA (Fig. 1): VFs are identified marking the vertebral body with six points to describe the vertebral shape and heights (Fig. 1). According to the quantitative morphometric approach, VFs are defined as mild, moderate, and severe based on a height ratio decrease of 20–25, 25–40, and more than 40%, respectively.[34] Guidelines also recommend the monitoring and correction of other potential risk factors for fractures, with special attention to hypogonadism, in addition to vitamin D deficiency, overtreatment of hypoadrenalism, and hypothyroidism.[20]

Arthropathy is one of the most invalidating complications of acromegaly that significantly impairs quality of life and affects both weight and non-weight bearing joints.[35] Early stage arthropathy is thought to be at least partially reversible by adequate treatment. However, when GH excess persists, the disease acquires features of a degenerative joint disease, resulting in scar, cyst, and osteophyte formation with further deterioration of joint architecture. In this late phase, changes become irreversible and may be GH-independent, and acromegaly treatment has only limited effects on joint symptoms.[35]

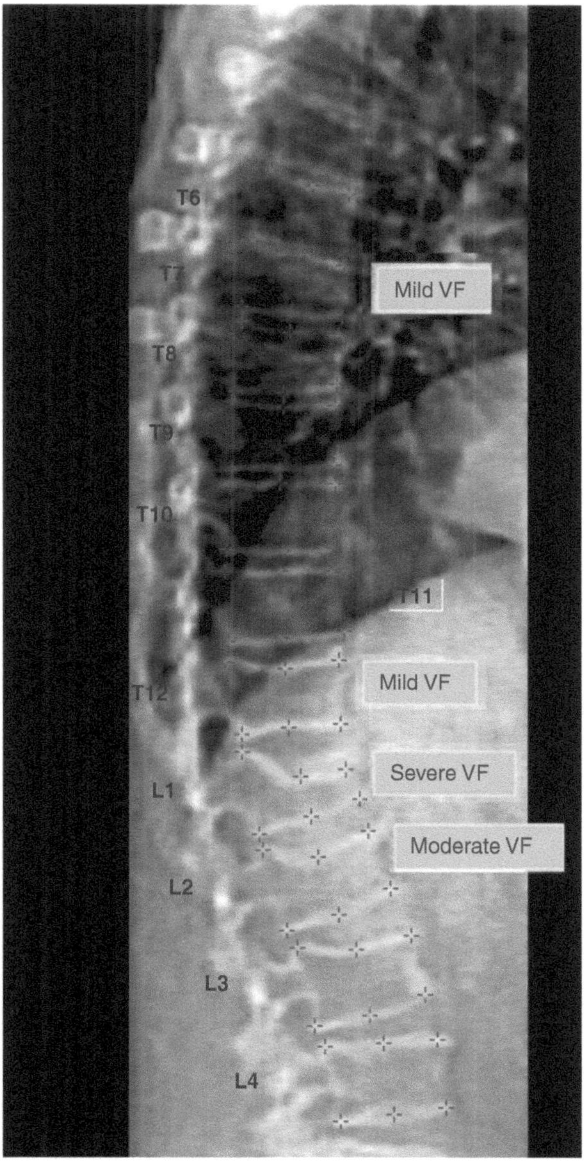

Figure 1 Lateral DXA image of thoracic and lumbar spine showing four VFs in an acromegaly patient (T7 and T12 mild, L2 moderate, L1 severe; according to Genant classification; see text for details); the fractures were identified by marking the vertebral bodies with six points to define the vertebral shape and heights.

It has been proposed that excess GH, with its promitogenic effect, might predispose to colorectal cancer. The increased rate of colon polyps, however, is not associated with increased mortality. Screening colonoscopy must be performed at diagnosis and, if negative, patients should follow the screening programs predisposed for the general population.[20] Instead, if IGF-I remains persistently over the upper normal range, or if the first colonoscopy is abnormal, follow-up should be more aggressive, in accordance with available clinical guidelines.

Acromegaly is associated with an increased mortality risk, although figures reported in recent studies seem to be quite lower than those initially reported, likely as expression of an improved management of disease. Mortality is closely related to GH hypersecretion.[2,3,36] However, the risk of mortality is also correlated with cardiovascular and neoplastic complications of acromegaly. Literature data seem to suggest that last follow-up serum GH is the strongest predictor of survival in the acromegalic population being the level of 2–2.5 g/L (measured by conventional and no more commonly used radioimmunoassays) a quite consistent cut-off above which an increase in mortality rate is observed.[36] Slightly more controversial is the role of IGF-I as predictor of mortality since in some studies this parameter did not show a strong association with mortality, while in other epidemiologic studies, elevated IGF-I was correlated with increased risk of mortality in patients with acromegaly.[2,3]

3. THERAPEUTIC OPTIONS

Available treatments for acromegaly include surgical removal of the pituitary adenoma, medical treatment, and radiation therapy with a stepwise approach.

Surgical approach is the only therapeutic option potentially leading, in the hands of an experienced neurosurgeon, to a rapid cure of acromegaly by removing GH-secreting adenoma.[37] However, in more than 50% of patients harboring a macroadenoma (which frequently, about 70% of cases, occurs in acromegaly due to delayed diagnosis), persistent GH hypersecretion after neurosurgery may be observed.[37] In this context and according to current guidelines, medical therapy has an important role in the management of acromegaly, either as first-line therapy in patients with a low probability of surgical cure or as adjuvant option in patients with persistent GH hypersecretion after unsuccessful neurosurgery.

3.1 First Generation SRLs

So far, five different somatostatin receptor subtypes (SSTRs) have been cloned. Human pituitary adenomas express mainly SSTR2 and SSTR5, which play a pathophysiologic role in mediating the inhibitory action of somatostatin on GH secretion. Besides the effects on GH hypersecretion, somatostatin can also inhibit cell growth and differentiation by stimulation of phosphotyrosine phosphatase or mitogen-activated protein kinase activity[38]

Indeed, the use of natural somatostatin in clinical practice is limited due to its short half-life (less than 3 min), with need of intravenous infusion and postinfusion rebound hormone hypersecretion. Synthetic SRLs with longer half-lives have therefore been developed for therapeutic uses. Octreotide long-acting repeatable (LAR) and lanreotide autogel (ATG) are the two long-acting SRLs currently approved for treatment of acromegaly, with different pharmacokinetic profiles but similar effectiveness.[39–41] These analogs, which are generally well tolerated, can be administered once a month. It is worthy to be mentioned that octreotide and lanreotide specifically target SSTR2. Lanreotide and octreotide differ in their mode of administration (long-acting lanreotide formulations are available in ready-to-use prefilled syringes that are injected subcutaneously; long-acting octreotide formulations require reconstitution before being injected intramuscularly), which might influence patient adherence to treatment.[42]

3.1.1 Effects on Biochemical Outcomes

In pivotal trials, more than two-third of patients treated with octreotide LAR or lanreotide ATG achieved biochemical control of acromegaly.[39,41] However, in observational studies involving nonselected cohort of patients, the percentages of success with these drugs resulted to be lower than those previously reported in the predesigned trials. In fact, from available databases biochemical control of acromegaly could be obtained in less than 50% of patients treated with SRLs,[43] likely due to the inclusion of unselected patients, not long enough duration of therapy or inadequate drug dosing, and a lower rate of treatment persistence or adherence outside the setting of controlled trials. Over the recent years, there has been convincing evidence that the effectiveness of SRLs may be significantly improved by increasing the SRL dose over the conventional schedules[44,45] or performing a surgical debulking of tumor mass before using SRLs.[46] In fact, 40 mg every 4 weeks[44] and more recently 60 mg every 4 weeks of octreotide LAR[45] were demonstrated to be effective in normalizing IGF-I in several patients only partially responsive to SRLs at conventional doses. Based on this evidence, the current guidelines[5] recommend increasing the SRL monthly dose before switching to other therapeutic options in patients partially responding to conventional doses of SRLs.

3.1.2 Effects on Tumor Mass and Clinical Outcomes

Despite the relatively modest shrinkage effect observed in the first clinical trials with regular ocreotide, more recent studies with real long-acting SRLs, such as octreotide LAR and lanreotide ATG, have shown that tumor shrinkage occurs as class effect in more than 50% of patients with acromegaly,[16,17] with an average volume reduction of about 50% versus pretreatment.[16] Such an entity of tumor shrinkage reveals that the effects on tumor mass of SRLs could be more than a simple radiological phenomenon, especially in patients with macroadenoma in whom the absolute size decrease could be more relevant in terms of relieving compression of surrounding structures.[47] Proportion of patients experiencing tumor shrinkage is higher when SRLs are used as first-line therapy, whereas tumor size does not seem to be a reliable predictor of tumor shrinkage since microadenomas and macroadenomas showed comparable response to SRLs.[16] The shrinkage effect of SRLs weakly correlates with duration of therapy; in fact shrinkage may occur in a number of patients after relatively short-term treatment.[48] All these findings may be important when SRL therapy is proposed as neoadjuvant therapy in patients with macroadenomas in order to induce tumor shrinkage and improve outcome of subsequent neurosurgical intervention.[48,49] The shrinkage effects of SRLs are not sustained since when the drug is discontinued, tumor regrowth may occur within the first 6 months and this event is important when patients on SRLs are switched to pegvisomant (see Section 2.4). Biochemical response has also been investigated as a determinant of tumor shrinkage, but data on this issue are also controversial. Prevalence of tumor shrinkage was shown to be higher in patients achieving either "safe" GH levels or normalization of IGF-I,[16] but it was reported to occur even in patients who did not achieve complete biochemical control under SRL treatment.[16–18] This latter evidence may be a rationale for combining SRLs and pegvisomant in patients who did not achieve biochemical control by SRLs alone.[50]

SRLs were shown to improve early almost all clinical symptoms of acromegaly. Effects on chronic complications are variable in relationship with age of patients, duration of disease, coexistence of risk factors, and mainly biochemical control of acromegaly. The improvement of some clinical end-points may occur early during the first months of treatment with

SRLs,[51] providing a further evidence to support the use of these drugs as first line therapy of acromegaly.

SRLs have marginal effects on blood pressure in acromegaly.[27] Although specific studies on the treatment of hypertension in acromegaly are lacking, patients should be treated according to local or international guidelines. Interestingly, epithelial sodium channel blockers, such as amiloride, may be of some efficacy in controlling hypertension caused by GH hypersecretion.[52] SRLs were shown to improve diastolic and systolic heart function and to decrease left-ventricle hypertrophy.[27] Interestingly, the beneficial effects of SRLs on cardiomiopathy were shown to occur even in patients without biochemical control of acromegaly,[53] consistently with the hypothesis that SRLs may have direct effects on peripheral targets not entirely related to inhibition of GH hypersecretion. However, when cardiomiopathy is either long-standing or affecting older patients, the effects of SRLs are marginal and specific cardiologic treatment should be implemented. SRLs improve sleep apnea syndrome by reduction of soft tissue swelling and decreasing chemosensitivity to hypoxia, usually within the first months of therapy, although a complete resolution of the syndrome is not generally achieved in most patients even after a long-standing control of acromegaly.[54] Therefore, every effort should be made to improve compliance with prescribed treatments including continuous positive air pressure device, and consultation with maxillofacial surgeons is advised and elective surgery should be undertaken, if needed.[20] Arthropathy is another complication which does not usually resolve by biochemical control of acromegaly. In fact, although several ultrasonography studies show partial reversibility of joint space thickening after a period of SRLs therapy, the characteristic radiographic phenotype of acromegalic arthropathy persists after achievement of biochemical disease control in the majority of patients, reflecting the irreversibility of the joint disease.[20,35]

The effects of SRLs on glucose metabolism have been extensively studied over the recent years. Somatostatin and its analogs are able to inhibit insulin secretion and this effect may potentially impair glucose homeostasis in acromegaly especially in patients with pre-existing glucose intolerance or type 2 diabetes.[43] Indeed, a recent meta-analysis evidenced that currently available SRLs, lanreotide ATG, and octreotide LAR, may have an overall marginal clinical impact on glucose homeostasis in this setting,[55] even when used at high doses.[56] However, a clinically significant deleterious glycometabolic effect may be observed in some patients treated with these drugs, especially when acromegaly is not biochemically controlled by the treatment.[56]

3.2 New SRLs

Pasireotide is a multireceptor-targeted SRL derived from incorporation of four synthetic and two essential amino acids of somatostatin in the form of a novel basic *trans*-(L)-hydroxyproline aminoethyl urethane extension, phenyglycine, O-benzyl-tyrosine, and D-Trp to corresponding positions into a stable cyclohexapeptide. One of the peculiar properties of this molecule is its 20–30 times higher binding affinity to SSTR1, and a 40–100 times higher binding affinity to SSTR5, as compared to octreotide and lanreotide.[57]

An earlier proof-of-concept study showed that single doses of sc pasireotide (100 and 250 mcg) suppressed GH and IGF-I levels in a dose-dependent manner in 12 patients with acromegaly, and in 3 of them the pasireotide-induced GH suppression was greater than that obtained with octreotide.[58] In subsequent clinical trials, pasireotide was shown to normalize serum IGF-I values in 60% of acromegaly patients.[59,60] More recently, a long-acting pasireotide formulation was shown to have superior efficacy compared with octreotide LAR or lanreotide ATG,[61,62] providing a rationale for proposing this new drug to patients with acromegaly who are inadequately controlled using first-generation SRLs.

Notwithstanding the different spectrum of receptor affinity, pasireotide and first-generation SRLs showed comparable tumor shrinkage effects.[57] However, due to its higher affinity for SSTR5, pasireotide may potentially produce more detrimental effects on glucose homeostasis with respect to currently available SRLs.[61,62] In fact, SSTR5 is expressed by pancreatic beta-cells and modulates insulin secretion.[63] Studies performed in healthy volunteers showed that pasireotide inhibited insulin secretion and incretin response, with minimal inhibition of glucagon secretion and no impact on insulin sensitivity.[64] Hyperglycemia secondary to pasireotide should be treated in accordance to the last ADA and EASD treatment guidelines for type 2 diabetes mellitus,[65] including dietary modifications and exercise. In patients with marked insulin resistance, metformin should be the first-line therapy, unless contraindicated or not tolerated. If the glycemic goals are not reached with metformin alone, combination therapy is recommended: because insulin secretion and incretin response are reduced, the introduction of a GLP-1-based therapy can be useful. In general, patients treated with pasireotide and hyperglycemia not controlled by metformin can be started with an oral DPP-IV inhibitor (which has low hypoglycemic potential together with neutral effect on body weight) and, in a few cases, switched to injective GLP-1 analogs.[64,65] Sulfonylureas and the PPARγ agonist pioglitazone are theoretically alternative options; however, sulfonylurea use is associated with the risk of hypoglycemia whereas PPARγ agonists were shown to be associated with bone fractures (particularly important side effect in the setting of acromegaly).[65,66] Finally, insulin therapy might be necessary, starting with a once-daily application of a long-acting basal insulin analog and, if not sufficient, with multiple prandial insulin injections.[65]

3.3 Dopamine Agonists

GH-secreting cells harbor dopamine receptor-2 and in acromegaly a paradoxical suppression of GH secretion is induced by dopamine agonists.[67] Cabergoline has been used for the treatment of acromegaly with variable results.[68–70] A critical analysis of the literature data revealed that cabergoline is effective in achieving biochemical control of acromegaly in patients with mild GH hypersecretion and with serum IGF-I values lower than two times the upper limit of normal.[68] The doses of cabergoline in acromegaly are generally higher than those used in prolactinomas and results might be assessed with a short-term (3–6 months) trial with dose escalation from 1.5 to 3.5 mg per week if well tolerated. To improve the biochemical effectiveness, combination therapy with SRLs was proposed.[5] Cabergoline was shown to induce tumor shrinkage in some acromegaly patients, especially in those with adenomas cosecreting prolactin and GH.[68] The effects of cabergoline on chronic complications are largely unknown, although there are reassuring data on its cardiovascular safety[71] and a beneficial effect on metabolic complications might be expected considering the favorable effects of dopamine on glucose homeostasis.[72]

3.4 Pegvisomant

Pegvisomant is a 199 amino acid recombinant competitive GH antagonist and the mutation on Gly120Arg generates the antagonism. European Medicines Agency (EMA) licensed it for the treatment of acromegaly more than 10 years ago. It is able to efficaciously block IGF-I production by antagonizing peripheral GH actions, via inhibition of dimerization of the GH receptor.[73] It is pegylated in order to guarantee a more stable molecule, reducing its immunogenicity, delaying renal clearance, and so prolonging its half-life to approximately 100 h. Pegvisomant should be injected daily in order to obtain a fully effective therapy and the achievement of higher serum drug concentrations. The starting dose should usually be 10 mg/day, reaching a maximum maintenance dose of 30 mg/daily.[74]

3.4.1 Biochemical Outcomes

In order to evaluate the efficacy of this drug, serum GH levels are not useful not only because endogenous GH secretion may be elevated during treatment, due to negative feedback, but also because of the cross-reaction between GH and pegvisomant in most of the GH assays. This is why GH measurement is not recommended in monitoring treatment efficacy.[7] Therefore, normal IGF-I levels represent the most relevant parameter during pegvisomant treatment.[7] The reported effectiveness in normalizing IGF-I levels widely varies in different type of studies. While clinical trials have shown disease control in up to 95% of patients during several years of treatment,[75,76] the real life experiences have demonstrated lower successful rates, probably caused by a selection bias similar to that of SRL studies, since the existence of a true biochemical resistance has not been documented so far.[77,78]

Effectiveness of pegvisomant may be inversely related to baseline IGF-I, while male gender, leanness, and previous irradiation are positive prognostic features.[74] Actually, the role of d3 GHR polymorphism, which increases receptor sensitivity to GH, is still controversial.[79]

3.4.2 Tumor Mass and Clinical Outcomes

Although cases of tumor growth during pegvisomant therapy have been reported, in most of them specific adenoma biology seems to play a major role (i.e., pegvisomant is not causing adenoma growth *per se* but does not have any shrinkage effects on the adenoma since its mechanism of action is not central but peripheral).[80] Pituitary MRI should be performed every 6 months for the first year of treatment and then if no tumor regrowth is observed, interval between MRIs may increase to 12 months. Previous irradiation and shorter duration of pre-pegvisomant SRL therapy were associated with augmented risk of tumor growth.[80]

Significant improvement in comorbidities, such as regression of soft tissue swelling, excessive perspiration, asthenia, and cardiovascular complications have been reported so far.[74] Improvements have been observed in insulin sensitivity and, in several patients, in fasting and postprandial glycemia, as well as in HbA1c.[81,82] Long-term treatment induces a significant cardiac mass reduction as well as improves diastolic and systolic functions and exerts positive effects on arrhythmias, hyperkinetic syndrome, and blood pressure, in particular on diastolic values.[74] It has been demonstrated that all these improvements lower the predicted cardiovascular risk calculated with the Framingham risk score.[83] Regarding skeletal abnormalities, pegvisomant is able to normalize the increased bone turnover and ameliorate bone mineral density, also after short-term treatments.[84]

Pegvisomant should be considered in all patients resistant to SRLs[74] and could be also administered together with SRLs,[5,85] which allows the use of lower doses of a costly drug with preserved efficacy and possibly improved safety, due to the protective effects of SRLs on tumor growth.[50] Moreover, relevant results have been obtained so far treating acromegaly associated with McCune–Albright syndrome.[86] In fact, pegvisomant has been proposed as first-line medical therapy for these peculiar patients.[74]

Lipodystrophy may occur at the injection site, due to local lipolytic GH inhibition, so it is very important to rotate the injection site using the entire abdomen area in order to avoid not only lipodystrophy, but also frequent mild, erythematous skin reactions.[74] About 5% of subjects treated with pegvisomant develop elevation of serum transaminases within the first year of treatment, which sometimes are up to threefold over the upper normal limit, causing drug discontinuation. It is mandatory to check liver function before starting the medication, frequently during the first 6 months of therapy and then less frequently (3 to 6 months).[74] A common polymorphism found in Gilbert's syndrome was associated with pegvisomant liver injury, so pegvisomant should not be administered in these patients, as well as in all subjects with liver dysfunctions. Combination of SRLs and pegvisomant may increase the frequency of liver enzyme elevation.

4. CONCLUSIONS

Surgical, pharmacologic, and radiotherapic approaches are used to treat acromegaly. Consensus statement and guidelines on medical management of acromegaly and its chronic complications have been recently updated. First generation SRLs, octreotide LAR, and lanreotide autogel ATG, are the first-line medical therapy either in naive patients with a low probability of surgical cure or as adjuvant option in patients with GH hypersecretion after unsuccessful neurosurgery. These drugs are generally well tolerated and are effective in normalizing GH hypersecretion and IGF-I values, controlling tumor growth, and also improving signs and symptoms of comorbidities in several patients. However, criteria for disease control may not be fulfilled in a considerable proportion of patients undergoing medical treatment with octreotide and lanreotide even when administered at higher doses than the conventional schedules. Accordingly, these acromegaly patients require the administration of GH antagonist pegvisomant, which was shown to normalize serum IGF-I values and improve chronic complications of acromegaly in a large number of patients with poor response to octreotide and lanreotide. A new generation multiligand SRL, pasireotide, has been developed and recently approved for treatment of acromegaly. Although this drug was shown to be more effective in normalizing GH hypersecretion as compared to octreotide and lanreotide, its position in the therapeutic guidelines of acromegaly is yet to be defined.

REFERENCES

1. Melmed S. Medical progress: acromegaly. *N Engl J Med.* 2006;355:2558–2573.
2. Holdaway IM, Bolland MJ, Gamble GD. A meta-analysis of the effect of lowering serum levels of GH and IGF-I on mortality in acromegaly. *Eur J Endocrinol.* 2008;159:89–95.
3. Dekkers OM, Biermasz NR, Pereira AM, Romijn JA, Vandenbroucke JP. Mortality in acromegaly: a metaanalysis. *J Clin Endocrinol Metab.* 2008;93:61–67.
4. Melmed S, Colao A, Barkan A, Molitch M, Grossman AB, Kleinberg D, Clemmons D, Chanson P, Laws E, Schlechte J, Vance ML, Ho KKY, Giustina A. Acromegaly Consensus Group.Guidelines for acromegaly management: an update. *J Clin Endocrinol Metab.* 2009;94:1509–1517.
5. Giustina A, Chanson P, Kleinberg D, Bronstein MD, Clemmons DR, Klibanski A, Van der Lely AJ, Strasburger CJ, Lamberts SW, Ho KKY, Casanueva FF, Melmed S. A consensus on the medical treatment of acromegaly. *Nat Rev Endocrinol.* 2014;10:243–248.
6. Giustina A, Barkan A, Casanueva FF, Cavagnini F, Frohman L, Ho KKY, Veldhuis J, Wass J, Von Werder K, Melmed S. Criteria for cure of acromegaly: a consensus statement. *J Clin Endocrinol Metab.* 2000;85:526–529.
7. Giustina A, Chanson P, Bronstein MD, Klibanski A, Lamberts S, Casanueva FF, Trainer P, Ghigo E, Ho KKY, Melmed S. A consensus on criteria for cure of acromegaly. *J Clin Endocrinol Metab.* 2010;95:3141–3148.
8. Giustina A, Bronstein MD, Casanueva FF, Chanson P, Ghigo E, Ho KKY, Klibanski A, Lamberts S, Trainer P, Melmed S. Current management practices for acromegaly: an international survey. *Pituitary.* 2011;14:125–133.
9. Carmichael JD, Bonert VS, Mirocha JM, Melmed S. The utility of oral glucose tolerance testing for diagnosis and assessment of treatment outcomes in 166 patients with acromegaly. *J Clin Endocrinol Metab.* 2009;94:523–527.
10. Alexopoulou O, Bex M, Abs R, T'Sjoen G, Velkeniers B, Maiter D. Divergence between growth hormone and insulin-like growth factor-1 concentrations in the follow-up of acromegaly. *J Clin Endocrinol Metab.* 2008;93:1324–1330.
11. Freda PU. Monitoring of acromegaly: what should be performed when GH and IGF-I levels are discrepant? *Clin Endocrinol (Oxf).* 2009;71:166–170.
12. Bianchi A, Giustina A, Cimino V, Pola R, Angelini F, Pontecorvi A, De Marinis L. Influence of growth hormone receptor d3 and full-length isoforms on biochemical treatment outcomes in acromegaly. *J Clin Endocrinol Metab.* 2009;94:2015–2022.
13. Trainer PJ, Barth J, Sturgeon C, Wieringaon G. Consensus statement on the standardisation of GH assays. *Eur J Endocrinol.* 2006;155:1–2.
14. Arafat AM, Möhlig M, Weickert MO, Perschel FH, Purschwitz J, Spranger J, Strasburger CJ, Schöfl C, Pfeiffer AF. Growth hormone response during oral glucose tolerance test: the impact of assay method on the estimation of reference values in patients with acromegaly and in healthy controls, and the role of gender, age, and body mass index. *J Clin Endocrinol Metab.* 2008;93:1254–1262.

15. Colao A, Auriemma RS, Lombardi G, Pivonello R. Resistance to somatostatin analogs in acromegaly. *Endocr Rev.* 2011;32:247–271.
16. Giustina A, Mazziotti G, Torri V, Spinello M, Floriani I, Melmed S. Meta-analysis on the effects of octreotide on tumor mass in acromegaly. *PLoS One.* 2012;7:e36411.
17. Mazziotti G, Giustina A. Effects of lanreotide SR and Autogel on tumor mass in patients with acromegaly: a systematic review. *Pituitary.* 2010;13:60–67.
18. Amato G, Mazziotti G, Rotondi M, Iorio S, Doga M, Sorvillo F, Manganella G, Di Salle F, Giustina A, Carella C. Long-term effects of lanreotide SR and octreotide LAR on tumour shrinkage and GH hypersecretion in patients with previously untreated acromegaly. *Clin Endocrinol (Oxf).* 2002;56:65–71.
19. Giustina A, Casanueva FF, Cavagnini F, Chanson P, Clemmons D, Frohman LA, Gaillard R, Ho KKY, Jaquet P, Kleinberg DL, Lamberts SW, Lombardi G, Sheppard M, Strasburger CJ, Vance ML, Wass JA, Melmed S. Diagnosis and treatment of acromegaly complications. *J Endocrinol Invest.* 2003;26:1242–1247.
20. Melmed S, Casanueva FF, Klibanski A, Bronstein MD, Chanson P, Lamberts SW, Strasburger CJ, Wass JAH, Giustina A. A consensus on the diagnosis and treatment of acromegaly complications. *Pituitary.* 2013;16:294–302.
21. Møller N, Jørgensen OL. Effects of growth hormone on glucose, lipid and protein metabolism in human subjects. *Endocr Rev.* 2009;30:152–177.
22. Kasayama S, Otsuki M, Takagi M, Saito H, Sumitani S, Kouhara H, Koga M, Saitoh Y, Ohnishi T, Arita N. Impaired beta-cell function in the presence of reduced insulin sensitivity determines glucose tolerance status in acromegalic patients. *Clin Endocrinol (Oxf).* 2000;52:549–555.
23. Rosario PW. Frequency of acromegaly in adults with diabetes or glucose intolerance and estimated prevalence in the general population. *Pituitary.* 2011;14:217–221.
24. American Diabetes Association.Standards of medical care in diabetes—2014. *Diabetes Care.* 2014;37(Suppl 1):S14–S80.
25. Boero L, Manavela M, Meroño T, Maidana P, Gómez Rosso L, Brites F. GH levels and insulin sensitivity are differently associated with biomarkers of cardiovascular disease in active acromegaly. *Clin Endocrinol (Oxf).* 2012;77:579–585.
26. Tan KC, Shiu SW, Janus ED, Lam KS. LDL subfractions in acromegaly: relation to growth hormone and insulin-like growth factor-I. *Atherosclerosis.* 1997;129:59–65.
27. Maison P, Tropeano AI, Macquin-Mavier I, Giustina A, Chanson P. Impact of somatostatin analogs on the heart in acromegaly: a metaanalysis. *J Clin Endocrinol Metab.* 2007;92: 1743–1747.
28. De Menis E, Giustina A, Colao A, Degli Uberti E, Ghigo E, Minuto F, Bogazzi F, Drigo R, Cattaneo A, Aimaretti G. COM.E.T.A. Italian Study Group.Assessment of the awareness and management of sleep apnea syndrome in acromegaly. The COM.E.TA (Comorbidities Evaluation and Treatment in Acromegaly) Italian Study Group. *J Endocrinol Invest.* 2011;34:60–64Erratum in: *J Endocrinol Invest.* 2011; 34:169.
29. Attal P, Chanson P. Endocrine aspects of obstructive sleep apnea. *J Clin Endocrinol Metab.* 2010;95:483–495.
30. Mazziotti G, Biagioli E, Maffezzoni F, Spinello M, Serra V, Maroldi R, Floriani I, Giustina A. Bone turnover, bone mineral density and fracture risk in acromegaly: a meta-analysis. *J Clin Endocrinol Metab.* 2015;100:384–394.
31. Bonadonna S, Mazziotti G, Nuzzo M, Bianchi A, Fusco A, De Marinis L, Giustina A. Increased prevalence of radiological spinal deformities in active acromegaly: a cross-sectional study in postmenopausal women. *J Bone Miner Res.* 2005;20:1837–1844.
32. Mazziotti G, Bianchi A, Bonadonna S, Cimino V, Patelli I, Fusco A, Pontecorvi A, De Marinis L, Giustina A. Prevalence of vertebral fractures in men with acromegaly. *J Clin Endocrinol Metab.* 2008;93:4649–4655.

33. Mazziotti G, Bianchi A, Porcelli T, Mormando M, Maffezzoni F, Cristiano A, Giampietro A, De Marinis L, Giustina A. Vertebral fractures in patients with acromegaly: a 3-year prospective study. *J Clin Endocrinol Metab.* 2013;98:3402–3410.
34. Genant HK, Jergas M, Palermo L, et al. Comparison of semiquantitative visual and quantitative morphometric assessment of prevalent and incident vertebral fractures in osteoporosis. The Study of Osteoporotic Fractures Research Group. *J Bone Miner Res.* 1996;11:984–996.
35. Claessen KM, Mazziotti G, Biermasz NR, Giustina A. Bone and joint disorders in acromegaly. *Neuroendocrinology.* 2015. DOI: 10.1159/000375450; PMID: 25633971. [epub ahead of print].
36. Ayuk J, Clayton RN, Holder G, Sheppard MC, Stewart PM, Bates AS. Growth hormone and pituitary radiotherapy, but not serum insulin-like growth factor-I concentrations, predict excess mortality in patients with acromegaly. *J Clin Endocrinol Metab.* 2004;89:1613–1617.
37. Abu Dabrh AM, Mohammed K, Asi N, Farah WH, Wang Z, Farah MH, Prokop LJ, Katznelson L, Murad MH. Surgical interventions and medical treatments in treatment-naive patients with acromegaly: systematic review and meta-analysis. *J Clin Endocrinol Metab.* 2014;99:4003–4014.
38. Eigler T, Ben-Shlomo A. Somatostatin system: molecular mechanisms regulating anterior pituitary hormones. *J Mol Endocrinol.* 2014;53:R1–R19.
39. Murray RD, Melmed S. A critical analysis of clinically available somatostatin analog formulations for therapy of acromegaly. *J Clin Endocrinol Metab.* 2008;93:2957–2968.
40. Chanson P, Boerlin V, Ajzenberg C, Bachelot Y, Benito P, Bringer J, Caron P, Charbonnel B, Cortet C, Delemer B, Escobar-Jiménez F, Foubert L, Gaztambide S, Jockenhoevel F, Kuhn JM, Leclere J, Lorcy Y, Perlemuter L, Prestele H, Roger P, Rohmer V, Santen R, Sassolas G, Scherbaum WA, Schopohl J, Torres E, Varela C, Villamil F, Webb SM. Comparison of octreotide acetate LAR and lanreotide SR in patients with acromegaly. *Clin Endocrinol (Oxf).* 2000;53:577–586.
41. Carmichael JD, Bonert VS, Nuño M, Ly D, Melmed S. Acromegaly clinical trial methodology impact on reported biochemical efficacy rates of somatostatin receptor ligand treatments: a meta-analysis. *J Clin Endocrinol Metab.* 2014;99:1825–1833.
42. Salvatori R, Nachtigall LB, Cook DM, Bonert V, Molitch ME, Blethen S, Chang S. Effectiveness of self- or partner-administration of an extended-release aqueous-gel formulation of lanreotide in lanreotide-naive patients with acromegaly. *Pituitary.* 2010;13: 115–122.
43. Giustina A, Karamouzis I, Patelli I, Mazziotti G. Octreotide for acromegaly treatment: a reappraisal. *Expert Opin Pharmacother.* 2013;14:2433–2447.
44. Colao A, Pivonello R, Auriemma RS, Galdiero M, Savastano S, Lombardi G. Beneficial effect of dose escalation of octreotide-LAR as first-line therapy in patients with acromegaly. *Eur J Endocrinol.* 2007;157:579–587.
45. Giustina A, Bonadonna S, Bugari G, Colao A, Cozzi R, Cannavo S, De Marinis L, Degli Uberti E, Bogazzi F, Mazziotti G, Minuto F, Montino M, Ghigo E. High-dose intramuscular octreotide in patients with acromegaly inadequately controlled on conventional somatostatin analogue therapy: a randomized controlled trial. *Eur J Endocrinol.* 2009;161: 331–338.
46. Colao A, Attanasio R, Pivonello R, Cappabianca P, Cavallo LM, Lodrini A, Lombardi G, Cozzi R. Partial surgical removal of growth hormone secreting pituitary tumors enhances the response to somatostatin analogs in acromegaly. *J Clin Endocrinol Metab.* 2006;91:85–92.
47. Melmed S, Sternberg R, Cook D, Klibanski A, Chanson P, Bonert V, Vance ML, Rhew D, Kleinberg D, Barkan A. A critical analysis of pituitary tumor shrinkage during primary medical therapy in acromegaly. *J Clin Endocrinol Metab.* 2005;90:4405–4410.

48. Caron PJ, Bevan JS, Petersenn S, Flanagan D, Tabarin A, Prévost G, Maisonobe P, Clermont A. PRIMARYS Investigators.Tumor shrinkage with lanreotide Autogel 120 mg as primary therapy in acromegaly: results of a prospective multicenter clinical trial. *J Clin Endocrinol Metab.* 2014;99:1282–1290.
49. Carlsen SM, Lund-Johansen M, Schreiner T, Aanderud S, Johannesen O, Svartberg J, Cooper JG, Hald JK, Fougner SL, Bollerslev J. Preoperative octreotide treatment in newly diagnosed acromegalic patients with macroadenomas increases cure short-term postoperative rates: a prospective, randomized trial. *J Clin Endocrinol Metab.* 2008;93: 2984–2990.
50. Gola M, Bonadonna S, Mazziotti G, Amato G, Giustina A. Resistance to somatostatin analogs in acromegaly: an evolving concept? *J Endocrinol Invest.* 2006;29:86–93.
51. Annamalai AK, Webb A, Kandasamy N, Elkhawad M, Moir S, Khan F, Maki-Petaja K, Gayton EL, Strey CH, O'Toole S, Ariyaratnam S, Halsall DJ, Chaudhry AN, Berman L, Scoffings DJ, Antoun NM, Dutka DP, Wilkinson IB, Shneerson JM, Pickard JD, Simpson HL, Gurnell M. A comprehensive study of clinical, biochemical, radiological, vascular, cardiac, and sleep parameters in an unselected cohort of patients with acromegaly undergoing presurgical somatostatin receptor ligand therapy. *J Clin Endocrinol Metab.* 2013;98: 1040–1050.
52. Kamenický P, Mazziotti G, Lombès M, Giustina A, Chanson P. Growth hormone, insulin-like growth factor-1, and the kidney: pathophysiological and clinical implications. *Endocr Rev.* 2014;35:234–281.
53. De Marinis L, Bianchi A, Mazziotti G, Mettimano M, Milardi D, Fusco A, Cimino V, Maira G, Pontecorvi A, Giustina A. The long-term cardiovascular outcome of different GH-lowering treatments in acromegaly. *Pituitary.* 2008;11:13–20.
54. Davi MV, Dalle Carbonare L, Giustina A, Ferrari M, Frigo A, Lo Cascio V, Francia G. Sleep apnoea syndrome is highly prevalent in acromegaly and only partially reversible after biochemical control of the disease. *Eur J Endocrinol.* 2008;159:533–540.
55. Mazziotti G, Floriani I, Bonadonna S, Torri V. Chanson P, Giustina A. Effects of somatostatin analogs on glucose homeostasis: a metaanalysis of acromegaly studies. *J Clin Endocrinol Metab.* 2009;94:1500–1508.
56. Mazziotti G, Porcelli T, Bogazzi F, Bugari G, Cannavò S, Colao A, Cozzi R, De Marinis L, Degli Uberti E, Grottoli S, Minuto F, Montini M, Spinello M, Giustina A. Effects of high-dose octreotide LAR on glucose metabolism in patients with acromegaly inadequately controlled by conventional somatostatin analog therapy. *Eur J Endocrinol.* 2011;164:341–347.
57. Giustina A, Mazziotti G, Maffezzoni F, Amoroso V, Berruti A. Investigational drugs targeting somatostatin receptors for treatment of acromegaly and neuroendocrine tumors. *Expert Opin Investig Drugs.* 2014;23:1619–1635.
58. Petersenn S, Schopohl J, Barkan A, Mohideen P, Colao A, Abs R, Buchelt A, Ho YY, Hu K, Farrall AJ, Melmed S, Biller BM. Pasireotide Acromegaly Study Group.Pasireotide (SOM230) demonstrates efficacy and safety in patients with acromegaly: a randomized, multicenter, phase II trial. *J Clin Endocrinol Metab.* 2010;95:2781–2789.
59. Petersenn S, Farrall AJ, De Block C, Melmed S, Schopohl J, Caron P, Cuneo R, Kleinberg D, Colao A, Ruffin M, Hermosillo-Reséndiz K, Hughes G, Hu K, Barkan A. Long-term efficacy and safety of subcutaneous pasireotide in acromegaly: results from an open-ended, multicenter, Phase II extension study. *Pituitary.* 2013;14:132–140.
60. Petersenn S, Bollerslev J, Arafat AM, Schopohl J, Serri O, Katznelson L, Lasher J, Hughes G, Hu K, Shen G, Reséndiz KH, Giannone V, Beckers A. Pharmacokinetics, pharmacodynamics and safety of pasireotide LAR in patients with acromegaly: a randomized, multicenter, open-label, phase I study. *J Clin Pharmacol.* 2014;54:1308–1317.
61. Colao A, Bronstein MD, Freda P, Gu F, Shen CC, Gadelha M, Fleseriu M, Van der Lely AJ, Farrall AJ, Hermosillo-Reséndiz K, Ruffin M, Chen Y, Sheppard M. Pasireotide

versus octreotide in acromegaly: a head-to-head superiority study. *J Clin Endocrinol Metab.* 2014;99:791–799.

62. Gadelha MR, Bronstein MD, Brue T, Coculescu M, Fleseriu M, Guitelman M, Pronin V, Raverot G, Shimon I, Lievre KK, Fleck J, Aout M, Pedroncelli AM, Colao A. Pasireotide C2402 Study Group.Pasireotide versus continued treatment with octreotide or lanreotide in patients with inadequately controlled acromegaly (PAOLA): a randomized, phase 3 trial. *Lancet Diabetes Endocrinol.* 2014;11:875–884.
63. Dietrich H, Hu K, Ruffin M, Song D, Bouillaud E, Wang Y, Hasskarl J. Safety, tolerability, and pharmacokinetics of a single dose of pasireotide long-acting release in healthy volunteers: a single-center Phase I study. *Eur J Endocrinol.* 2012;166:821–828.
64. Henry RR, Ciaraldi TP, Armstrong D, Burke P, Ligueros-Saylan M, Mudaliar S. Hyperglycemia associated with pasireotide: results from a mechanistic study in healthy volunteers. *J Clin Endocrinol Metab.* 2013;98:3446–3453.
65. Inzucchi SE, Bergenstal RM, Buse JB, Diamant M, Ferrannini E, Nauck M, Peters AL, Tsapas A, Wender R, Matthews DR. Management of hyperglycemia in type 2 diabetes: a patient-centered approach. Position statement of the American Diabetes Association (ADA) and the European Association for the Study of Diabetes (EASD). *Diabetes Care.* 2012;35:1364–1379.
66. Mancini T, Mazziotti G, Doga M, Carpinteri R, Simetovic N, Vescovi PP, Giustina A. Vertebral fractures in males with type 2 diabetes treated with rosiglitazone. *Bone.* 2009;45:784–788.
67. Chiodini PG, Liuzzi A, Botalla L, Cremascoli G, Silvestrini F. Inhibitory effect of dopaminergic stimulation on GH release in acromegaly. *J Clin Endocrinol Metab.* 1974;38:200–206.
68. Sandret L, Maison P, Chanson P. Place of cabergoline in acromegaly: a meta-analysis. *J Clin Endocrinol Metab.* 2011;96:1327–1335.
69. Marazuela M, Ramos-Leví A, Sampedro-Núñez M, Bernabeu I. Cabergoline treatment in acromegaly: pros. *Endocrine.* 2014;46:215–219.
70. Kasuki L, Vieira Neto L, Gadelha MR. Cabergoline treatment in acromegaly: cons. *Endocrine.* 2014;46:220–225.
71. Maione L, Garcia C, Bouchachi A, Kallel N, Maison P, Salenave S, Young J, Assayag P, Chanson P. No evidence of a detrimental effect of cabergoline therapy on cardiac valves in patients with acromegaly. *J Clin Endocrinol Metab.* 2012;97:E1714–E1719.
72. Mazziotti G, Gazzaruso C, Giustina A. Diabetes in Cushing syndrome: basic and clinical aspects. *Trends Endocrinol Metab.* 2011;22:499–506.
73. Kopchick JJ, Parkinson C, Stevens EC, Trainer PJ. Growth hormone receptor antagonists: discovery, development, and use in patients with acromegaly. *Endocr Rev.* 2002;23: 623–646.
74. Giustina A, Ambrosio MR, Beck-Peccoz P, Bogazzi F, Cannavò S, De Marinis L, De Menis E, Grottoli S, Pivonello R. Use of pegvisomant in acromegaly. An Italian Society of Endocrinology guideline. *J Endocrinol Invest.* 2014;37:1017–1030.
75. Trainer PJ, Drake WM, Katznelson L, Freda PU, Herman-Bonert V, Van der Lely AJ, Dimaraki EV, Stewart PM, Friend KE, Vance ML, Besser GM, Scarlett JA, Thorner MO, Parkinson C, Klibanski A, Powell JS, Barkan AL, Sheppard MC, Malsonado M, Rose DR, Clemmons DR, Johannsson G, Bengtsson BA, Stavrou S, Kleinberg DL, Cook DM, Philips LS, Bidlingmaier M, Strasburger CJ, Hackett S, Zib K, Bennett WF, Davis RJ. Treatment of acromegaly with the growth hormone-receptor antagonist pegvisomant. *N Engl J Med.* 2000;342:171–177.
76. Van der Lely AJ, Hutson RK, Trainer PJ, Besser GM, Barkan AL, Katznelson L, Klibanski A, Herman-Bonert V, Melmed S, Vance ML, Freda PU, Stewart PM, Friend KE, Clemmons DR, Johannsson G, Stavrou S, Cook DM, Philips LS, Strasburger CJ, Hackett S, Zib KA, Davis RJ, Scarlett JA, Thorner MO. Long-term treatment of

acromegaly with pegvisomant, a growth hormone receptor antagonist. *Lancet.* 2001;358:1754–1759.
77. Grottoli S, Maffei P, Bogazzi F, Cannavò S, Colao A, Ghigo E, Gomez R, Graziano E, Monterubbianesi M, Jonsson P, De Marinis L. ACROSTUDY: the Italian experience. *Endocrine.* 2015;48:334–341.
78. Giustina A. Optimal use of pegvisomant in acromegaly: are we getting there? *Endocrine.* 2015;48:3–8.
79. Bianchi A, Mazziotti G, Tilaro L, Cimino V, Veltri F, Gaetani E, Pecorini G, Pontecorvi A, Giustina A, De Marinis L. Growth hormone receptor polymorphism and the effects of pegvisomant in acromegaly. *Pituitary.* 2009;12:196–199.
80. Van Der Lely AJ, Biller BM, Brue T, Buchfelder M, Ghigo E, Gomez R, Hey-Hadavi J, Lundgren F, Rajicic N, Strasburger CJ, Webb SM, Koltowska-Häggström M. Long-term safety of pegvisomant in patients with acromegaly: comprehensive review of 1288 subjects in ACROSTUDY. *J Clin Endocrinol Metab.* 2012;97:1589–1597.
81. Barkan AL, Burman P, Clemmons DR, Drake WM, Gagel RF, Harris PE, Trainer PJ, Van Der Lely AJ, Vance ML. Glucose homeostasis and safety in patients with acromegaly converted from long-acting octreotide to pegvisomant. *J Clin Endocrinol Metab.* 2005;90:5684–5691.
82. De Marinis L, Bianchi A, Fusco A, Cimino V, Mormando M, Tilaro L, Mazziotti G, Pontecorvi A, Giustina A. Long-term effects of the combination of pegvisomant with somatostatin analogs (SSA) on glucose homeostasis in non-diabetic patients with active acromegaly partially resistant to SSA. *Pituitary.* 2007;10:227–232.
83. Berg C, Petersenn S, Lahner H, Herrmann BL, Buchfelder M, Droste M, Stalla GK, Strasburger CJ, Roggenbuck U, Lehmann N, Moebus S, Jöckel KH, Möhlenkamp S, Erbel R, Saller B, Mann K. Investigative Group of the Heinz Nixdorf Recall Study and the German Pegvisomant Observational Study Board and Investigators.Cardiovascular risk factors in patients with uncontrolled and long-term acromegaly: comparison with matched data from the general population and the effect of disease control. *J Clin Endocrinol Metab.* 2010;95:3648–3656.
84. Parkinson C, Kassem M, Heickendorff L, Flyvbjerg A, Trainer PJ. Pegvisomant-induced serum insulin-like growth factor-I normalization in patients with acromegaly returns elevated markers of bone turnover to normal. *J Clin Endocrinol Metab.* 2003;88: 5650–5655.
85. Neggers SJ, Van Aken MO, Janssen JA, Feelders RA, De Herde WW, Van Der Lely AJ. Long-term efficacy and safety of combined treatment of somatostatin analogs and pegvisomant in acromegaly. *J Clin Endocrinol Metab.* 2007;92:4598–4601.
86. Salenave S, Boyce AM, Collins MT, Chanson P. Acromegaly and McCune–Albright syndrome. *J Clin Endocrinol Metab.* 2014;99:1955–1969.

CHAPTER FIVE

Genetic Predictors of Response to Different Medical Therapies in Acromegaly

Ana M. Ramos-Leví*, Ignacio Bernabeu†, Miguel Sampedro-Núñez*, Mónica Marazuela*,1

* Department of Endocrinology and Nutrition, Hospital Universitario la Princesa, Instituto de Investigación Princesa, Universidad Autónoma de Madrid, Madrid, Spain

† Department of Endocrinology and Nutrition, Complejo Hospitalario Universitario de Santiago de Compostela, Servicio Gallego de Salud (SERGAS); Universidad de Santiago de Compostela, La Coruña, Spain

[1] Corresponding author: e-mail: monica.marazuela@salud.madrid.org.

Contents

Abstract

In the era of predictive medicine, management of diseases is evolving into a more personal and individualized approach, as more data are available regarding clinical, biochemical, radiological, molecular, histopathological, and genetic aspects. In the particular setting of acromegaly, which is a rare, chronic, debilitating, and disfiguring disease, an optimized approach deems even more necessary, especially because of an associated increased morbidity and mortality, the impact on patients' quality of life, and the increased cost of frequently necessary life-long treatments. In this paper, we

Progress in Molecular Biology and Translational Science, Volume 138
ISSN 1877-1173
http://dx.doi.org/10.1016/bs.pmbts.2015.10.016

review the available studies that address potential genetic influences on acromegaly, their role in the outcome, and response to treatments, as well as their contribution to the risk of developing side effects. We focus mainly on pharmacogenetic factors involved during treatment with dopamine agonists, somatostatin analogs, and pegvisomant. Specifically, mutations in dopamine receptors, somatostatin receptors, growth hormone receptors, and metabolic pathways involved in growth hormone action; polymorphisms in the insulin-like growth factor and the insulin-like growth factor binding proteins; and polymorphisms in other genes that may determine differences in the frequency of developing adverse events.

1. INTRODUCTION

Acromegaly is a rare, chronic, debilitating, and disfiguring disease, which entails a wide range of cardiovascular, respiratory, endocrine, and metabolic morbidities, as well as an increased risk for all-cause mortality.[1,2] Transsphenoidal surgery is the first treatment of choice in most cases, and in patients with intrasellar microadenomas, surgical removal provides biochemical control with normalization of IGF-I in 75–95% of patients. However, control rates are lower (40–70%) in patients with macroadenomas, requiring the use of other adjuvant therapies, including medical options and/or radiotherapy, which aim to control both GH/IGF-I hypersecretion and residual tumor volume.[1]

The first medical approaches to acromegaly began in the 1970s with the use of dopamine agonists.[3–6] Nowadays, the more selective dopamine-2 receptor agonist cabergoline (CAB) may be a useful alternative in a subset of patients who exhibit modest elevations of IGF-I, or in the setting of combination therapy with somatostatin analogs (SSA) or pegvisomant (PEG).[7–10]

The development of SSA, which mimics the native physiologic inhibitor of GH secretion, set the beginning of a new era in the management of acromegaly. These drugs provided clinicians with an effective tool for acromegalic patients, since they allowed both an adequate control of GH/IGF-I hypersecretion, as well as a direct inhibitory effect on the somatotroph adenoma.[11–13] Consequently, SSA are considered as the first-line medical treatment option not only in the setting of persistent residual biochemical disease despite surgical intervention, but also as primary or preoperative treatment in some cases.[14,15] However, about one-third of patients treated with SSA exhibit partial or complete resistance and do not achieve biochemical normalization.[16,17]

The first and only available GH-receptor antagonist PEG competes with endogenous GH for its receptor and prevents functional dimerization and signal transduction by the GH receptor (GHR), therefore inhibiting its activation and subsequent actions.[18] In preclinical studies, PEG has demonstrated to be highly effective in normalizing IGF-I values (rates above 90%), including patients who are partially or completely resistant to other medical therapies.[19–21] In the clinical setting, studies from groups who are particularly interested in the management of acromegaly reported a biochemical efficacy of PEG of over 80%,[22,23] but this efficacy was only 63.2% in the multinational observational study Acrostudy, which included over 1200 acromegalic patients.[24] In Europe, PEG is indicated for patients who have undergone surgery and/or radiotherapy, and are resistant to SSA. Safety issues with PEG do not seem to be of concern, and adverse events potentially associated to this drug only seem to occur occasionally. Specifically, these side effects are mainly related to abnormal liver function test results, which rarely entails withdrawal of PEG, and development of lipohypertrophy at the injection site. Long-term studies have not demonstrated an increased rate of tumor growth or progression.[24–30]

Treatment of acromegaly is expensive,[31,32] especially regarding perioperative care and medical management, which will often need to be lifelong.[33] Therefore, cost issues should not be underestimated. All clinical, biochemical, radiological, molecular, and histopathological aspects should be taken into consideration in order to be able to effectively and wisely decide which drug is the best available one for a particular patient with acromegaly. In other words, we need tools that help clinicians develop therapeutic algorithms that consider a personalized and individual approach.[34] In this setting, pharmacogenomic studies, which identify genetic predictors of response to medical therapies in acromegaly, will surely help to optimize the management of patients, by helping in the choice of treatment, and potentially allowing a quicker normalization of biochemical parameters. Even though they are difficult to perform and evaluate due to the rarity of the disease, in the short-term, these studies will probably allow individualization of available treatments, thus providing a more efficient approach than current algorithms, which are mainly based upon "trial-and-error," with sequential additional treatment options provided when disease is not controlled.[34]

2. GENETIC FACTORS INVOLVED IN TREATMENT WITH DOPAMINE AGONISTS

In acromegaly, dopamine paradoxically inhibits GH secretion by binding to one of the five subtypes of dopamine receptors (D1–D5).[3,4] Dopamine receptors are grouped into two families according to their molecular, physiologic, and pharmacologic characteristics: the *D1-like*, which includes subtypes D1 and D5, and the *D2-like*, which includes subtypes D2, D3, and D4.[35] *D2-like* receptor subtypes inhibit adenilcyclase activation through their coupling to the G-protein; they further activate potassium channels and undergo several processes of homo- and heterodimerization with other G-protein-coupled receptors, such as somatostatin receptors (SSTRs).

Currently available dopamine agonists in clinical practice are all *D2-like*, and they may be ergot-derived, such as bromocriptine, lisuride, and CAB, or nonergot-derived, like quinagolide.

CAB is well tolerated when administered orally, and provides a good drug bioavailability. In prolactinomas, dopamine agonists inhibit prolactin synthesis and secretion and induce a reduction in tumor size causing perivascular fibrosis and cell necrosis.[5,36]

In the case of acromegaly, CAB has proven to be moderately effective, yet clinically relevant, in controlling hypersecretion and tumor proliferation in patients with acromegaly. In fact, this has been observed in studies using treatment schemes with CAB on its own,[9,37–39] as well as in combination regimes with SSA[9] or PEG.[40,41] Furthermore, complete remission of acromegaly has been described after prolonged long-term treatment.[42]

Data from observational studies have highlighted the efficacy of CAB in patients with persistent acromegaly who previously underwent surgery, especially in cases with moderate residual biochemical activity. The predictive value of concomitant hyperprolactinemia or positive prolactin immunohistochemistry of the tumor remain controversial, so current recommendations do not consider these factors as essential for considering or disregarding CAB as a treatment option for acromegaly.[43,44] Further molecular and histopathologic studies, which evaluate the expression of D1 and D2 receptors in somatotroph adenomas, will probably help clinicians choose CAB in a more rational way in the acromegaly medical treatment algorithm.

3. GENETIC FACTORS INVOLVED IN TREATMENT WITH SOMATOSTATIN ANALOGS

SSA are considered the cornerstone for the medical management of acromegaly. In fact, not only have they proven to be effective for disease control in patients with persistent acromegaly after surgery, but also as primary presurgical treatment, and as the only exclusive approach when surgery is contraindicated or cannot be performed.[14] Long-term follow-up studies have described GH and IGF-I normalization rates of 34–70%,[17,45,46] and tumor shrinkage of over 20% in up to 75% of cases.[14,47,48]

3.1 Somatostatin Receptors

SSAs exert their biologic actions by binding to a family of G-protein-coupled, seven transmembrane-spanning somatostatin receptors (designated sst 1–5) in somatotroph adenomas.[49,50] This leads to decreased hormonal secretion, decreased growth and proliferation rates, increased apoptosis, and/or inhibition of cell signaling and protein synthesis, including inhibition of production and secretion of various angiogenic factors. Somatostatin receptors are encoded by five separate, intronless genes (SSTR1–SSTR5), which have been classically considered to give rise to five different somatostatin receptors, named sst1 through sst5. However, recent studies have revealed the existence of new truncated, albeit functional sst5 variants, generated by noncanonical splicing, which in humans bear five and four transmembrane domains (instead of the usual seven), and are therefore referred to as sst5TMD4 and sst5TMD5. These truncated variants sst5TMD5 and sst5TMD4, which are absent in normal pituitary, have been demonstrated in pituitary tumors.[51,52]

As sst subtypes represent obligatory mediators of SSA actions, in acromegaly SSA are effective because somatotroph adenomas express sst receptors. Currently available SSAs, octreotide and lanreotide, both in their short- or long-acting forms, exert their actions mainly as sst2/sst5 agonists.[53,54] In this regard, variability in the sst1–5 expression profile in the pituitary has been suggested as a potential predictive factor for SSA response.[55] Specifically, a low expression or absence of SSTR2, an SSTR2/SSTR5 ratio <1.3[56] have all been associated with higher rates of resistance to SSA treatment. However, the response to SSA in acromegaly cannot be completely and exclusively explained by the pattern of expression of the classical sst receptors found in tumor samples.[57] In this regard, the sst5 variant, sst5TMD4, which can be found in pituitary cells in some somatotroph adenomas, is able to interact with sst2, leading to a disrupted signaling, and thus can reduce the response to SSA treatment in acromegalic patients.[51,52,58]

Tumor histopathology is also relevant for predicting both its behavior and its response to SSA.[59] According to electron-microscopy studies, GH-secreting adenomas can be classified into two distinct variants, reflecting different patterns of GH-containing secretory granules: densely or sparsely granulated. The former subtype is characterized by a high number of large

GH-containing secretory granules resembling nontumorous somatotrophs. On the other hand, sparsely granulated somatotropinomas, that is, those that stain less in immunocytochemical analysis, show only sparse and small secretory granules. Reports agree that these latter adenomas are usually larger, more invasive, and more resistant to SSA than densely granulated adenomas, mainly due to the different SSTR2 expression pattern.[60,61]

3.2 G-Protein-Linked Receptor Mutations

GTP-binding (G) proteins are a large family of proteins involved in coupling hormone and neurotransmitter receptors, and they are also regulated upon SSTR activation. Mutations causing a constitutive activation of this G-protein family generate proliferative signals and promote tumor growth. In this regard, activating mutations of the α-subunit of the G-protein coupled to the GHR (GSP mutations) have been associated to an increased cAMP activation, and thus to a better response to SSA.[16]

3.3 AIP and ZAC 1

Germline mutations in the AIP gene (*Aryl hydrocarbon receptor-interacting protein*) or in ZAC1 gene (also known as ZAC/PLAGL1/LOT-1, which is a tumor-suppressor gene encoding a zinc-finger protein known as ZAC1 that exerts an antiproliferative effect by inducing apoptosis and G1 cell-cycle arrest) have also been associated to differences in response to SSA. In fact, data derived from several reports suggest the involvement of ZAC1 in the mechanism of action of SSAs, and more recent findings suggest that SSAs could regulate AIP, which in turn can stimulate ZAC1, therefore suggesting the existence of a SSA–AIP–ZAC1–somatostatin effect pathway.[62,63] In this regard, AIP has been proposed as a relevant biomarker to predict the potential response of somatotroph adenomas to SSA treatment, especially if considered together with the tumor's SSTR expression profile.[64]

3.4 Other Molecules and Pathways

Further characterization of other molecules, such as, for example SHP1, SHP2, MAPK, or GSK3 β,[65] involved in the molecular activation cascade triggered by SSA binding (cAMP inhibition and protein kinase A and phosphate-phosphotyrosine activation) will surely help the development of specific target treatments.[63]

Moreover, the Ras-Raf-MEK-ERK1/2-p27 pathway, which is downstream from SSTRs, plays a role in hormone release, apoptosis, and cell proliferation, and is thus involved in the therapeutic response to SSA. When inhibition of the MAPK signaling is not effective enough, a poor disease prognosis has been observed in some cases of acromegaly.[34] Inhibition of RAS-1 kinase phosphorylation by nonphosphorylated RAK kinase inhibitory protein (RKIP) may attenuate MAPK signaling, and low levels of RKIP and the lack of RAF kinase inhibition have been associated with a worse response to SSA.[34]

Additionally, the development of new SSA-like pasireotide (SOM230), with a broader and higher affinity for all SSTR subtypes (except for SSTR4)[66–69] may open new perspectives for the management of patients with acromegaly resistant to currently available therapies. In this way, target therapies will allow an individualized approach to acromegalic patients, optimizing the efficacy of the management of this disease.

4. GENETIC FACTORS INVOLVED IN TREATMENT WITH PEGVISOMANT

4.1 GH Receptor

4.1.1 Mechanism of Action of Pegvisomant

GH interacts with its specific receptor, GHR, by means of which it exerts all its biologic actions. The GHR is a transmembrane protein, with three domains: an extracellular, a transmembrane, and an intracellular one. GHR is constitutively dimerized in the cell membrane and there may be certain polymorphisms, the majority of them due to variants in exons 3 and 9.[70,71] The gene encoding this receptor, the *GHR*, is found in the short arm of chromosome 5 (p13.1-p12).[72] This gene comprises 9 exons, numbered 2 through 10: exon 2 encodes a signal peptide; exons 3 through 7 encode the extracellular or ligand-binding domain, exon 8 encodes the transmembrane domain, and exons 9 and 10 encode the intracellular domain (Fig. 1).[73]

Native GH binds to preformed GHR dimers.[74,75] Subsequently, GH binding to sites 1 and 2 of the GHR dimer leads to conformational changes, which then promote transphosphorylation of both GHR and Janus-kinase 2 (Jak2).[74,75] The latter, on its side, causes recruitment and further phosphorylation of other signal transducers and transcription activating signals (such as STAT 1, 3, 5A, and 5B), allowing the formation of STAT homo- and heterodimers, which are then translocated to the cell nucleus, where gene transcription is promoted. This Jak2/STAT pathway represents the main

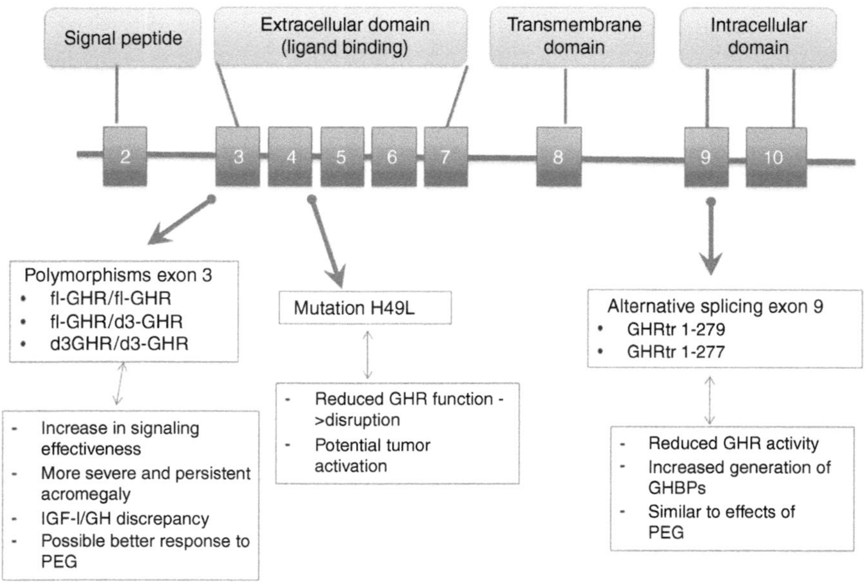

Figure 1 Schematic representation of the GHR, in the short arm of chromosome (p13.1-p12).[72] This gene comprises 9 exons, which encode the signal peptide, the extracellular domain, and the intracellular domain. Variants of the GHR: deletion of exon 3, H49L mutation, and truncated forms due to alternative splicing of exon 9.

signaling mechanism of GHR, which is necessary for regulation of the IGF-I gene transcription (STAT5). The other two Jak2-activated pathways comprise the mitogen-activated protein kinase (MAPK/ERK1/2) and the phosphoinositide-3 kinase (PI3K/AKT).[76–78] On its side, Jak2 activation promotes activation of the ubiquitin system, which leads to receptor conjugation and internalization. Generally, in normal conditions, GHR recycling is quick and only takes about 30–60 min, a time after which GHR is cleared by two mechanisms: proteolysis, to generate GH-binding proteins (GHBPs) and a continuous internalization process, which is dependent on ubiquitin, and will ultimately lead to endolysosomic degradation.[74]

PEG is analogous to GH, but exhibits changes in nine amino acids; one of them (G120K) confers its antagonist effect, while the remaining preserve their affinity for GHR binding. The molecule is further pegylated to extend its half-life.[18] Consequently, PEG binds to site 1 of the GHR, but is unable to bind to site 2, thus inhibiting the cascade that leads to Jak2 phosphorylation.[18] Therefore, PEG binds to and blocks available GHR, and competes with native GH. In addition, because of the absence of Jak2 phosphorylation, activation of the ubiquitin system (UbE) is also inhibited, contributing to slowing down the GHR internalization, which may explain the increase in GHBPs observed during PEG treatment.[79]

4.1.2 Genomic Deletion of Exon 3 in the Growth Hormone Receptor (d3-GHR)

Exon 3 of the GHR is flanked by two almost identical retroelements comprised of 251 bp, which are situated at 577 and 1821 bp from the exon, respectively. Some individuals exhibit a homologous recombination of these two retroelements, and produce 2.7-kb deletion, giving rise to two different isoforms of the GHRGHR: the full-length GHR, which retains the exon 3 (fl-GHR or fl/fl), and the exon-3-deleted GHR (d3-GHR), which is either homozygous (d3/d3) or heterozygous (fl/d3). Deletion of exon 3 produces a lack of amino acid residues 7–28 and the amino acid substitution A6D close to the N-terminus.[73] This induces subtle conformational changes in the extracellular domain, which do not affect the affinity, binding capacity, or internalization of both GHR isoforms, but which could facilitate ligand-induced activation of GHR (Fig. 1).[73,80] Around 50% of the Western population is homozygous for the fl-GHR, 30–40% is heterozygous (fl-GHR/d3GHR) and 10–20% is homozygous for the d3 variant.[73,80]

Preliminary studies did not consider this deletion to have a significant impact on the GHRs function.[81,82] However, subsequent reports demonstrated its association with a differential GH transduction.[80]

In vitro studies have demonstrated how transcriptional activity of homo- and heterodimers of the d3-GHR are around 30% higher than that of the fl-GHR homodimers.[80] In this regard, clinical studies in GH-deficient children carrying the d3-GHR variant have reported a better response to treatment with recombinant human GH.[80,83] In GH-deficient adults, however, some studies,[84,85] but not all,[86] have demonstrated this greater sensitivity to GH treatment in d3-GHR carriers.

Several studies, mostly multicenter, retrospective, and with a relatively small number of patients, have addressed the impact of the d3-GHR allele in acromegaly.[87–96] Patients carrying the d3-GHR variant exhibited a close relationship between serum GH and IGF-I concentrations; specifically, for a given IGF-I serum concentration, GH levels were lower in d3-GHR carriers.[87] Similarly, the presence of d3-GHR was associated to a higher frequency of biochemical discrepant results, with elevated IGF-I levels, but normal GH, both during SSA treatment, and after surgical intervention. In this regard, in a series of 84 patients with acromegaly, 70% of those who presented with biochemical discrepant results were d3-GHR allele carriers.[90]

In addition, in one of the largest series of acromegalic patients from a single center,[89] patients with heterodimers d3/fl-GHR, and especially those with homodimers d3/d3-GHR, presented a more active disease, higher rate of diabetes mellitus, and lower probability of achieving adequate biochemical control with either surgery, SSA and/or radiotherapy. Moreover, IGF-I levels were higher in homo- and heterozygous patients for d3-GHR, in comparison to patients with the fl-GHR allele. Furthermore, insufficient biochemical control was observed in 54% of patients who were fl/fl-GHR, 55% of those with d3/fl-GHR, and 77% of patients who were homozygous for the d3-GHR allele. This study suggested, in fact, that the best predictor of persistent elevated IGF-I levels during or after treatment was the presence of the d3-GHR allele, with even better predictive values than basal GH/IGF-I levels, age, or tumor size.[89]

The potential role of the d3-GHR in acromegaly and its outcome has been specifically addressed in several well-known studies.[89–91,94,97,98] None of them found a significant association between the presence of this allele and basal or biochemical characteristics, although subtle differences were found regarding weight, body mass index, and the percentage of macroadenomas in two of them.[91,94] Turning to the potential influence of this exon deletion in the development of comorbidities during long-term follow-up, some studies have described a greater prevalence of osteoarthropathy, vertebral fractures, dolicocolon, and colonic polyps, as well as a greater probability of maintaining normal tolerance to glucose, in carriers of the d3-GHR variant.[92,93,97]

Regarding response to treatment with PEG, in two studies,[91,94] where patients received PEG in monotherapy, carriers of the d3-GHR allele exhibited a better response: IGF-I normalization occurred earlier and the required PEG dose was 21–27% lower. Meanwhile, in the multicenter study performed by our group,[94] a multivariate analysis showed that presence of the d3-GHR allele and male sex were the only two relevant predictors of PEG dose necessary to achieve IGF-I normalization (Table 1).[30,94] Another multicenter study performed in Italy with 127 acromegalic patients (49% of whom were on combination therapy with SSA + PEG), found no association between the presence of the d3-GHR allele and response to treatment with PEG, either when used as monotherapy or in combination schemes with SSA.[95] However, given the observed deviation from Hardy–Weinberg equilibrium of the d3-GHR genotype distribution in their study, as well as in previous ones, they suggested an association of the d3-GHR variant with a phenotype characterized by resistance to traditional treatments and difficulties in disease control.[95]

Table 1 Factors Influencing Doses of Pegvisomant.

Factor	Effect
GH/IGF-I level	Direct correlation with PEG dose required
Sex	Higher doses required in men Higher GH and GHBP levels with lower IGF-I in women
Radiotherapy	Reduces PEG dose required
Weight	Increases PEG dose required
Type 2 diabetes mellitus	Increased insulin promotes increased hepatic GHR and therefore requires higher PEG doses
Combination therapy with SSA	Decreased insulin levels, and subsequent reduced GHR, so less PEG dose required
d3-GHR	Lower doses of PEG and fewer months to normalize IGF-I
Polymorphism promoter region IGF-I	Controversial reports

Abbreviations: GH, growth hormone; IGF-I, insulin-like growth factor; PEG, pegvisomant; d3-GHR, exon-3-deleted GH receptor; GHBP, growth hormone binding protein; SSA, somatostatin analogs. Adapted from Ref. [30].

To summarize, deletion of exon 3 of the GHR in patients with acromegaly seems to influence the clinical picture of the disease, its severity, and maybe the response to PEG. In this regard, blockage of the d3-GHR, which is functionally more active, may determine a greater effect than blockage of the native, fl/fl-GHR, which exhibits a lower functional activity. Possible explanations for the inconsistencies observed in results from different studies may concern the fact that the exon 3 deletion of the GHR gene still has a limited impact in clinical endocrinology,[99] or because studies have included a small and heterogeneous group of acromegalic patients. In any case, studies that specifically address the significance and clinical relevance of the d3-GHR in acromegaly deem necessary to open a new gate for targeted therapies based on pharmacogenomics, in order to personalize and individualize management of this rare disease.

4.1.3 Mutations in Exons 4 and 5 of the GHR

More than a decade ago, Asa *etal*, suggested that somatotroph cells exhibited an autonomous regulation system, which was based on GH-circulating levels and was mediated by GHR.[100] They examined the pituitary glands of giant transgenic mice expressing a GH agonist (E117L), dwarf transgenic mice expressing a GH antagonist (G119K), and dwarf mice devoid of the GHR/binding protein (GHR/BP). The authors observed a relationship between the rate of GH/GHR activity and pituitary histopathologic findings. They concluded that there was a synthetic and secretory hyperactivity in pituitary somatotrophs due to the reduced GH-feedback regulation, evidencing a direct GH feedback inhibition on pituitary somatotrophs, which may have implications for the use of GH analogs in different clinical settings.[100]

A couple of years later, the same research group published a study in which they evaluated mutations in the GHR in normal pituitaries and in 26 somatotropinomas, of which 12 were densely granulated and 14 sparsely granulated. The activating mutation Gsα for the GHRH receptor was found in five densely granulated adenomas, but in none of the sparsely granulated ones. However, six cases exhibited a heterozygous mutation in exon 4 of the GHR, which lead to a substitution in codon 49 of histidine for leucine in five cases and for arginine in one case (Fig. 1).

It is worth noting that exons 4 and 5 of the GHR, play a relevant role in allowing conformational and rotational changes in the extracellular domain of the GHR, which is necessary for GH binding.[18,75,101] Consequently, variants in the resulting amino acids may lead to changes in the affinity of GH–GHR binding.[71,102]

For instance, experimental studies comparing native GHR and mutant GHR-H49L proved that the latter exhibited a shorter half-life, a lower affinity for GH, lower transcriptional activity, and undergoes an abnormal posttranslational processing, which leads to accumulation of immature forms of native GHR in the cell membrane.[71,103] As a result, the GHR-H49L variant hypothetically originates a disruption in the physiologic role of the GHR and, subsequently, in GH control on somatotroph cells, as it has been previously observed.[100] In this regard, the H49L-GHR mutation has been suggested as a putative mediator in the pathogenesis of sparsely granulated adenomas, the ones with a more aggressive phenotype. Although not all adenomas with this pattern presented this mutation, it is worth remarking that they exhibited a marked reduction in the intensity of GHR immunoreactivity, suggesting the possible existence of other alternative mechanisms leading to disruption or instability of the GHR, and, subsequently, to a sparsely granulated pattern.[103–105] Asa *et al.* further evaluated this hypothesis by examining the effects of GHR interruption gene in GH4 somatotroph cell cultures and in eight primary human pituitary densely granulated somatotroph tumors exposed to increasing pharmacologic doses of PEG. After 4 days, examination of cultured cells revealed morphologic changes typical of sparsely granulated tumors, which pointed out that aggresome formation was a result of GHR antagonism in human somatotroph cells, and that it could be considered a surrogate marker of reduced GHR signaling, consistent with the effect of the GHR mutations identified in sparsely granulated pituitary somatotroph tumors. In other words, GHR disruption, either because of the H49L mutation, other yet-to-be identified mutations, or after PEG treatment, may directly promote cellular hyperactivity, increased GH secretion,[105] and hyperplasia,[100] conveying a relevant role in the pathogenesis of sparsely granulated somatotropinomas.[103,105] In this setting, the suggested hypothesis is that interruption of the GH–GHR axis in somatotroph cells from densely granulated pituitary adenomas with PEG could potentially promote morphologic and histopathologic changes leading to an "activation" of the tumor itself. In fact, this theory has also been remarked in subsequent reviews,[105,106] although it has not been definitively confirmed by other authors.

For the time being, there is sufficient data that rule out a significant increase in tumor growth during PEG treatment in comparison to long-term use of SSA.[25,107,108] In addition, extensive immunohistochemistry studies evaluating numerous markers (ACTH, FSH, LH, GH, TSH, PRL, cytokeratin 18, P16, P21, P27, P53, Bcl-2, Ki-67/MIB-1, galectin 3, vascular-endothelial growth factors VEGFR-1, VEGFR2, and VEGFR3, platelet-derived growth factor PDGF, and its receptor PDGFRα) have not proven significant differences in tumors before or after treatment with PEG.[25] Nevertheless, some authors have described a subtle relationship between tumor growth and higher GH levels during PEG treatment,[26] as well as between higher GH levels and the tumor's own insulin receptor.[25]

4.1.4 Alternative Splicing of Exon 9

Exons 9 and 10 of the *GHR* gene encode the intracellular domain of the GHR. As a result, alternative splicing of exon 9 may cause deletion of a fragment of 26 bp, leading to a stop codon at position 280, entailing loss of 97.5% of the intracellular domain (GHRtr 1–279).[109] Another truncated variant of the GHR occurs due to complete loss of exon 9, which encodes a receptor known as GHRtr 1–277 (Fig. 1).[110,111] The frequency and inter-individual variants regarding expression of these isoforms in tumor tissues has not been fully elucidated, nor the potential relevance of identifying other alternative splicing sites. The putative role of these truncated GHR on somatotroph physiology is yet to be investigated, although preliminary data point out that these variants share certain functional characteristics with native GHR blocked by PEG. Hence, an interesting line of future investigation will surely concern examination of the relationship between expression levels of truncated GHR, clinical aspects of acromegaly, and response to PEG treatment.

Absence and/or loss of a functional GHR intracellular domain entails a number of consequences. For instance, *in vivo* studies in mice[112] expressing truncations in GHR at residue m569 (loss of the C-terminal subdomain) and m391 (a broader loss of the intracellular domain, including box 1), result in reduced GH signaling of about 70–100%, lower levels of IGF-I transcripts (80–90%), and limited postnatal growth. Meanwhile, in humans, truncated GHR variants are functionally inactive and seem to act as negative regulators of the fl/GHR.[109,113–115] In addition, the subsequent aberrant internalization process that occurs[74] entails the generation of more GHBPs,[109,113–115] which correspond to the circulating soluble form of the extracellular domain of the GHR after its proteolytic rupture mediated by the enzyme TACE.[115–118] GHBPs compete with GH in their binding to GHR and inhibit it, as well as inhibiting IGF-I synthesis[115] and regulating transcription of the GHR gene by reducing it in the presence of high GHBP levels.[119,120] Estrogens, on their part, seem to modulate the expression of these truncated GHR variants.[121]

Binding of PEG to the GHR dimer originates a similar situation to that of truncated GHR. Specifically, the PEG-blocked dimer is internalized, but is not activated,[122,123] so it does not lead to phosphorylation of Jak2, nor activation of the UbE system.[124] Therefore, as it has already been pointed out, internalization of the PEG–GHR complex slows down and generates GHBPs,[79,124,125] which, in turn, downregulate the expression levels of GHR in the cell surface, originating a refractory period for GHR binding of about 2–3 h.[126] Although PEG binds to GHR with a much lower affinity than native GH, its affinity to GHBPs is similar,[126] which is one of the reasons that may explain the high doses of PEG required to effectively block the effects of GH.

In summary, polymorphisms in exon 9 may hypothetically influence the outcomes of PEG treatment through different mechanisms, but well-designed studies with consistent results have not yet been communicated.

4.1.5 Other GHR Polymorphisms

In a randomly selected subgroup of 25 acromegalic patients from a multicenter study, five missense single-nucleotide polymorphisms found in a public database, that is, rs348388342 (W4R), rs6413484 (V162I), rs6181 (R179H), rs35395580 (NM_000163.2:c.569_570insG), and rs6180 (I544L), were analyzed. Only rs6180 had a variability sufficient for statistical analysis, but no association with clinical, biochemical, or pharmacological parameters was found.[95]

4.2 IGF-I Polymorphisms

Polymorphisms in the promoter region of the IGF-I gene, which is encoded in chromosome 12q, has been recently described. These polymorphisms could hypothetically modify the clinical expression and the response to treatments in acromegaly.

The IGF-I gene polymorphism consists of a highly polymorphic microsatellite composed of a variable number of cyto-adenosine (CA) repeats (from 10 to 24), situated 1 kb upstream form the transcription site of IGF-I (Fig. 2). The most common allele in the Caucasian population contains 19CA repeats (192 bp).[127] Differences in the number of CA repeats have been associated with serum IGF-I levels, although controversial results have been communicated. Also, the number of CA repeats have been associated with various clinical conditions or risk situations, such as age-related IGF1 decline, final adult height, risk of developing diabetes mellitus and myocardial infarction, survival after a myocardial infarction in diabetic patients, efficacy of recombinant human GH (rhGH) treatment in GH-deficient children and Turner syndrome and cancer susceptibility.[127–132] However, results have not been fully consistent across studies.

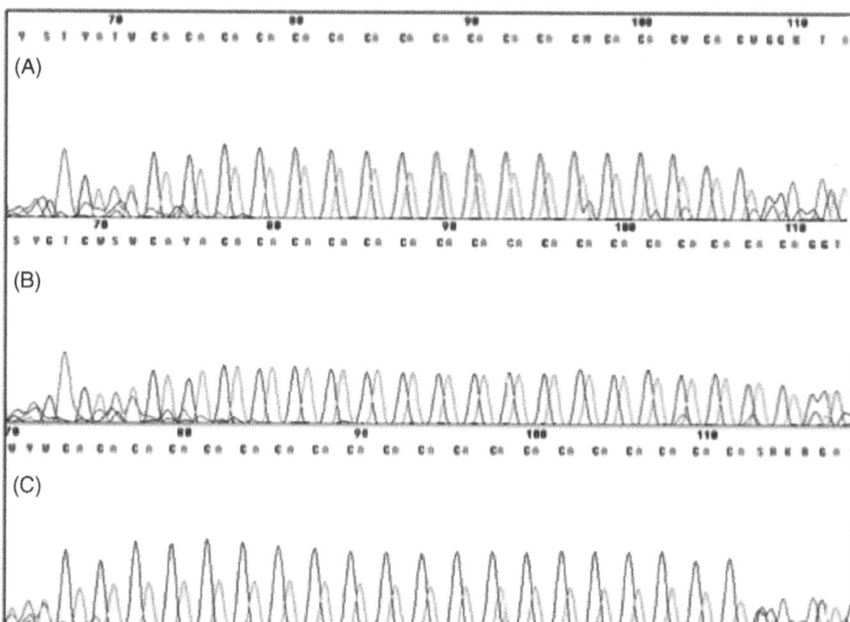

Figure 2 IGF-I(CA) polymorphism. Examples from patients with (A) 18 CA repeats, (B) 19 CA repeats, and (C) 20 CA repeats.

In children, on the other hand, this polymorphism does not seem to be determinant in the response to rhGH treatment, since no differences have been observed in final adult height in relation to IGF-I gene variants.[133,134]

A previous publication analyzed the influence of this IGF-I polymorphism in a sample of 34 acromegalic patients.[135] The authors observed that those with the 194 bp genotype (i.e., 20 CA repeats) exhibited larger, more invasive tumors, and required adjuvant postsurgical treatments more frequently.

We recently performed a multicenter study in 124 patients with acromegaly, for whom genotypes of the IGF-I promoter region polymorphisms were available in 113 patients.[136] We observed that the genotype frequencies were in Hardy–Weinberg equilibrium: 15 patients (13.2%) had <192 bp (<19 CA repeats), 82 (72.6%) had 192/194 bp (19/20 CA repeats), and 16 (14.2%) had >194 bp (>20 CA repeats). This meant that 44 patients (39.0%) presented the wild-type alleles IGF(CA)19 (192/192 bp) and 69 (61.0%) did not (192/− or −/−). We did not find an association between the number of CA repeats and patients' baseline characteristics. Nevertheless, a trend for higher nadir GH values was observed in patients with <19 CA repeats. Regarding its potential influence on treatment outcomes, our results did not portray differences in response to treatments or adverse effects according to the genetic variant of the IGF-I promoter region polymorphism, suggesting that genotype variants of the IGF(CA) polymorphism may have only a minor influence on the treatment outcome. Despite the limitations of this study, this is biologically plausible, since current treatments for acromegaly mainly act via the SSTRs and GHR pathways. However, we considered our findings as being clinically relevant, since knowledge of the patients' specific polymorphic genotype may help predict specific phenotypic associations.[136]

4.3 IGFBP-3 Polymorphisms

The vast majority of circulating IGF-I binds to IGF-specific binding proteins (IGFBPs), which modify its bioavailability by extending its half-life and modulating its biologic actions on target tissues. IGFBP3 is the most abundant circulating subtype, which transports more than 75% of serum IGFs, as well as modulates its circulating levels.

Similar to what was hypothesized regarding polymorphisms in the promoter region of the IGF-I, polymorphisms in the promoter region of the IGFBP3 gene, which is encoded in chromosome 7p, have been suggested to influence disease phenotype and treatment outcome in acromegaly.[137]

Recent research has revealed the presence of five single-nucleotide polymorphisms (SNPs) in the IGFBP3 promoter region, of which the most relevant was an A to C nucleotide change located 202 bp upstream to the transcription site (−202A/C). The promoter activity is significantly higher in A allele carriers, entailing a correlation between IGFBP3 circulating levels and the number of A alleles at this site. In this regard, IGFBP3 levels would be higher in AA, and decreasing in a step-wise manner in AC and CC. Data derived from studies performed in twins evidenced that 60% or interindividual variability in IGFBP3 levels was genetically determined.[138] In fact, genetic variations in the IGFBP3 gene have been associated to final adult height and response to rhGH treatment in GH-deficient children, small for gestational age and Turner's syndrome.[139–141]

In the specific setting of acromegaly, data is still scarce. A previous report of 25 acromegalic patients[135] did not find an association between variants of this polymorphism and clinical or biochemical characteristics, nor with the outcome or adverse events of treatments. On the contrary, in our study,[136] baseline index IGF-I differed between the three groups (3.1 ± 1.5% in CC, 2.1 ± 1.0% in AC, and 2.7 ± 1.1% in AA). We hypothesize that the AA genotype would determine higher IGFBP3 levels and, consequently, a lower bioavailable-free IGF-I and a higher stimulation of GH secretion by the feedback mechanism of the somatotropic axis, which is usually preserved in patients with somatotropinomas. However, regarding response to treatments or adverse effects, our results were similar to that of the IGF(CA) polymorphism, in that it seemed that there was only a minor influence, similar to what others reported.[135,136]

4.4 Pegvisomant-Induced Liver Dysfunction and Polymorphisms of the UGT1A1*28 and Alcohol Dehydrogenase

Liver dysfunction is the most feared systemic adverse event potentially induced by PEG treatment.[142] The German observational study, in which all patients were included at the start of treatment with PEG, reported its occurrence in 8.4% of cases.[143] More recently, the observational multinational study Acrostudy[24] included 1288 patients, of whom more than 80% were enrolled when they had already been on PEG therapy for a mean period of 1.3 years. In this case, the prevalence of liver dysfunction, reported as elevation of liver enzymes at least three times the upper limit of normal (ULN) was 2.5%. There have been some communications of acute hepatitis,[143–146] but in any case permanent liver damage has been reported.

In fact, in the majority of cases, the effect was limited to mild liver enzyme elevation, which rarely required withdrawal of treatment.[24]

Liver injury during PEG treatment has been reported to occur more frequently in patients with previously known liver disease, when PEG was associated to SSA treatment, and in patients with diabetes mellitus.[147] In this regard, it is considered an idiosyncratic effect that cannot be, therefore, predicted.[143]

Uridine-diphosphate-5′-glucoronosylstransferase (UGT) proteins are encoded by 15 polymorphic genes and originate two families depending on their sequence homology: UGT2 and UGT1A. All UGTs exhibit tissue expression patterns and bind several ligands, which then conjugate with uridine-diphospho-glucuronic acid to subsequently facilitate their biliary or renal excretion.[148–151] Specifically, the UGT1A1 family is involved in bilirubin conjugation and, till date, there have been over 113 UGT1A1 variants reported.[150]

Gilbert's syndrome is characterized by mild-unconjugated nonhemolytic hyperbilirubinemia, which does not lead to hepatic inflammation, fibrosis, chronic liver disease, or liver failure, so it is usually considered as a benign condition.[150] This syndrome was linked to a genetic variant of the human UGT1A1, UGT1A1*28, found in approximately 40% of Caucasian individuals, which consist of two additional bases (TA) in the sequence TATAA in the 5′ gene promoter region (Fig. 3).[152] This mutation leads to a reduction of around 70% of the gene's transcriptional activity, in comparison to native UGT1A1, and consequently results in a reduced liver glucuronidation

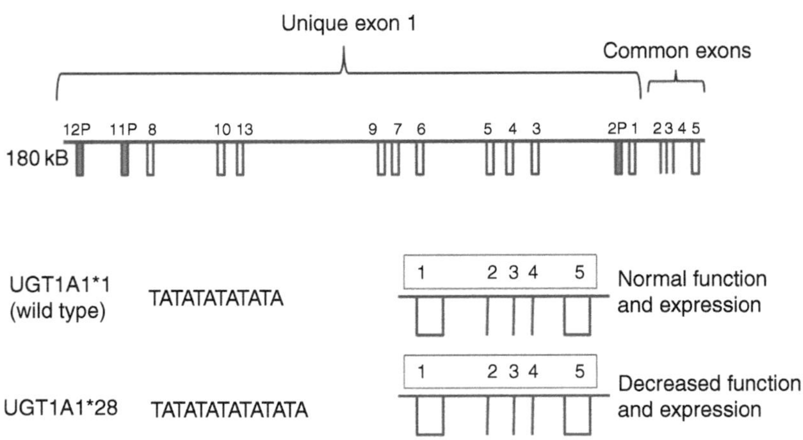

Figure 3 Schematic representation of the UGT1A1 polymorphism.

activity.[150] Such a reduction in the liver's ability to conjugate affects not only bilirubin metabolism, but also impacts drug therapy due to altered metabolism and excretion.[148,150,151,153,154] The estimated prevalence of this polymorphism in Caucasian population is around 15% for homozygous and up to 50% for heterozygous; and its presence is a necessary condition, albeit insufficient on its own, for the development of Gilbert's syndrome.[152]

PEG's excretion mechanism is not fully understood, but it probably involves a number of catabolic reactions, leading to the generation of pegylated residues, which are subsequently metabolized by the alcohol dehydrogenase (ADH). These latter pegylated residues may potentially develop liver and kidney toxicity in experimental animals, and, in fact, PEG-associated nephrotoxicity has been described.[155]

In this setting, and due to the acknowledgment of previous publications reporting the development of acute hepatitis during PEG treatment in a patient with Gilbert's syndrome,[143] our group investigated the influence of the UGT1A1*28 mutation in the development of severe liver injury, expressed as liver enzyme elevation 3 the ULN, during PEG monotherapy.[156] Of the 36 patients with active acromegaly who were resistant to SSA treatment and were on PEG monotherapy, we detected liver function test (LFT) abnormalities in 10 (28%). This high prevalence was probably due to the complete liver evaluation scheduled in all follow-up visits, as well as a systematic record of all LFT abnormalities, including isolated or transient changes, while previous observational studies had only reported clinically significant hepatic adverse events.

The prevalence of the UGT1A1*28 homozygous and heterozygous genotypes in acromegalic patients was 14 and 44%, respectively. Carriers of the UGT1A1*28 polymorphism had a higher incidence of LFT abnormalities than the UGT1A1 wild type (43% carriers vs. 7% wild type, $p = 0.024$); in fact, 9 out of 10 cases with LFT abnormalities occurred in carriers of the UGT1A1*28 polymorphism. This difference persisted when adjusted in an all-factors multiple regression analysis for age, gender, alcohol consumption, and UGT1A1*28 polymorphism. Furthermore, a stepwise multivariate likelihood binary logistic regression analysis identified male gender and UGT1A1*28 polymorphism as the only significant predictors for the development of LFT abnormalities during PEG monotherapy.[156]

A recent multicenter, retrospective, observational study of 112 patients with active acromegaly who were resistant to SSA treatment, and in whom PEG was used in either monotherapy or combination with SSA, evaluated the possible associations between PEG-induced liver injury and the

UGT1A1*28 variant, and additionally with some ADH polymorphisms involved in liver disorders or alcohol metabolism.[157] 6.3% of patients had previous abnormal LFT, and 39.3% had current or past cholelithiasis or biliary sand. During PEG therapy, 17% of patients developed abnormal LFT to some extent, that is, elevation above the ULN of at least one of the liver enzymes, in at least one of the follow-up visits. Overt hepatotoxicity, with liver enzyme elevation of at least three times the ULN, developed in 5 (4.5%) patients; of these, 3 were men, 4 had previously diagnosed gallstones, 2 were on combination treatment with SSA, 3 had evidence of previous LFT abnormalities in clinical history, and all received at least one well-known hepatotoxic drug (in fact, 2 patients took 12–13 oral medications). Patients (54.4%) carried the UGT1A1*28 polymorphic allele in either homo- or heterozygosis. Logistic and linear regression analyses showed an association between LFT abnormalities during follow-up visits, prior events of LFT abnormalities in medical history, and the number of concomitant medications, other than SSA. However, no correlation between LFT alterations and UGT1A1*28 as well as ADH1C and B polymorphisms was found. The authors concluded that the UGT1A1*28 and ADH1C and B polymorphisms do not predict increased risk of hepatotoxicity during PEG therapy.[157]

In summary, the potential role of different genetic variants of the UGT system in the development of liver abnormalities during PEG therapy in patients with active acromegaly still needs to be elucidated. Further, specific and long-term studies exploring other polymorphisms, such as UGT1A7* and UGT1A6*2, could contribute to a better understanding.

5. CONCLUSIONS

There is still insufficient data regarding pharmacogenetic studies that evaluate the potential influence of genetic polymorphisms on acromegaly and its outcome. The main underlying reasons concern the rarity of this disease and the difficulty to perform randomized studies.

Till date, observational studies suggest that certain genetic variants can indeed predict the response to individual therapies. Current medical guidelines and algorithms for the management of acromegaly follow a sequential "step-by-step" basis, and medical treatment frequently needs to be maintained life-long, entailing an important economic burden. Therefore, every clinical, biochemical, radiological, molecular, histopathological, and genetic aspect that may be considered should help us to optimize an effective

Table 2 Summary of the Main Clinical, Biochemical, Radiological, Molecular, Histopathological, and Genetic Aspects That May be Considered in Order to Optimize the Management of a Patient With Active Acromegaly.

Potential Influence	Clinical and Molecular Factors
Increased response to dopamine agonists	D1 and D2 expression
Resistance to SSA treatment	Sparsely granulated adenomas
	Low expression of SSTR2
	Low SSTR2/SSTR5 ratio
	Presence of SSTR5TMD4
	Positive *AIP* mutation
	Low AIP/ZAC1 expression
	High RAF kinase activity
	Lack of *gsp* mutation
Increased PEG doses required	Absence of d3-GHR
	Mutations in exons 4 and 5
	Alternative splicing in exon 9
Other potential influencing factors	IGFBP3 polymorphisms
	IGF-I polymorphisms
	UGT polymorphisms

Abbreviations: D1 and D2, dopamine receptors 1 and 2; SSA, somatostatin analogs; SSTR, somatostatin receptor; AIP, aryl hydrocarbon receptor-interacting protein; d3-GHR, exon 3 deletion in the GH receptor; IGF-I, insulin-like growth factor; PEG, pegvisomant.

approach to the patient with active acromegaly (Table 2). In other words, clinicians should aim at managing this disease, as any other, in a personalized and individual way. And in this setting, pharmacogenomic studies, which identify genetic predictors of response to medical therapies in acromegaly, will surely help, by elucidating the best choice of treatment and potentially allowing a quicker normalization of biochemical parameters.

REFERENCES

1. Melmed S. Medical progress: acromegaly. *N Engl J Med.* 2006;355(24):2558–2573.
2. Holdaway IM, Rajasoorya RC, Gamble GD. Factors influencing mortality in acromegaly. *J Clin Endocrinol Metab.* 2004;89(2):667–674.
3. Liuzzi A, Chiodini PG, Botalla L, Cremascoli G, Silvestrini F. Inhibitory effect of L-Dopa on GH release in acromegalic patients. *J Clin Endocrinol Metab.* 1972;35(6):941–943.
4. Mims RB, Stein RB, Bethune JE. The effect of a single dose of L-dopa on pituitary hormones in acromegaly, obesity, and in normal subjects. *J Clin Endocrinol Metab.* 1973;37(1):34–39.
5. Thorner MO, Chait A, Aitken M, Benker G, Bloom SR, Mortimer CH, et al. Bromocriptine treatment of acromegaly. *Br Med J.* 1975;1(5953):299–303.
6. Wass JA, Thorner MO, Morris DV, Rees LH, Mason AS, Jones AE, et al. Long-term treatment of acromegaly with bromocriptine. *Br Med J.* 1977;1(6065):875–878.

7. Higham CE, Atkinson AB, Aylwin S, Bidlingmaier M, Drake WM, Lewis A, et al. Effective combination treatment with cabergoline and low-dose pegvisomant in active acromegaly: a prospective clinical trial. *J Clin Endocrinol Metab.* 2012;97:1187–1193.
8. Bernabeu I, Alvarez-Escola C, Paniagua AE, Lucas T, Pavon I, Cabezas-Agrıicola JM, et al. Pegvisomant and cabergoline combination therapy in acromegaly. *Pituitary.* 2012;16:101–108.
9. Sandret L, Maison P, Chanson P. Place of cabergoline in acromegaly: a meta-analysis. *J Clin Endocrinol Metab.* 2011;96(5):1327–1335.
10. Abs R, Verhelst J, Maiter D, Van Acker K, Nobels F, Coolens JL, et al. Cabergoline in the treatment of acromegaly: a study in 64 patients. *J Clin Endocrinol Metab.* 1998;83(2): 374–378.
11. Plewe G, Beyer J, Krause U, Neufeld M, del Pozo E. Long-acting and selective suppression of growth hormone secretion by somatostatin analogue SMS 201-995 in acromegaly. *Lancet.* 1984;2(8406):782–784.
12. Ch'ng LJ, Sandler LM, Kraenzlin ME, Burrin JM, Joplin GF, Bloom SR. Long term treatment of acromegaly with a long acting analogue of somatostatin. *Br Med J (Clin Res Ed).* 1985;290(6464):284–285.
13. Lamberts SW, Oosterom R, Neufeld M, del Pozo E. The somatostatin analog SMS 201-995 induces long-acting inhibition of growth hormone secretion without rebound hypersecretion in acromegalic patients. *J Clin Endocrinol Metab.* 1985;60(6):1161–1165.
14. Melmed S, Colao A, Barkan A, Molitch M, Grossman AB, Kleinberg D, et al. Guidelines for acromegaly management: an update. *J Clin Endocrinol Metab.* 2009;94: 1509–1517.
15. Katznelson L, Laws Jr ER, Melmed S, Molitch ME, Murad MH, Utz A, et al. Acromegaly: an endocrine society clinical practice guideline. *J Clin Endocrinol Metab.* 2014;99(11):3933–3951.
16. Colao A, Auriemma RS, Lombardi G, Pivonello R. Resistance to somatostatin analogs in acromegaly. *Endocr Rev.* 2011;32(2):247–271.
17. Carmichael JD, Bonert VS, Nuno M, Ly D, Melmed S. Acromegaly clinical trial methodology impact on reported biochemical efficacy rates of somatostatin receptor ligand treatments: a meta-analysis. *J Clin Endocrinol Metab.* 2014;99(5):1825–1833.
18. Kopchick JJ, Parkinson C, Stevens EC, Trainer PJ. Growth hormone receptor antagonists: discovery, development, and use in patients with acromegaly. *Endocr Rev.* 2002;23 (5):623–646.
19. van der Lely AJ, Hutson RK, Trainer PJ, Besser GM, Barkan AL, Katznelson L, et al. Long-term treatment of acromegaly with pegvisomant, a growth hormone receptor antagonist. *Lancet.* 2001;358(9295):1754–1759.
20. Trainer PJ, Drake WM, Katznelson L, Freda PU, Herman-Bonert V, van der Lely AJ, et al. Treatment of acromegaly with the growth hormone-receptor antagonist pegvisomant. *N Engl J Med.* 2000;342(16):1171–1177.
21. Herman-Bonert VS, Zib K, Scarlett JA, Melmed S. Growth hormone receptor antagonist therapy in acromegalic patients resistant to somatostatin analogs. *J Clin Endocrinol Metab.* 2000;85(8):2958–2961.
22. Higham CE, Chung TT, Lawrance J, Drake WM, Trainer PJ. Long-term experience of pegvisomant therapy as a treatment for acromegaly. *Clin Endocrinol (Oxf).* 2009;71(1):86–91.
23. Marazuela M, Lucas T, Alvarez-Escola C, Puig-Domingo M, de la Torre NG, de Miguel-Novoa P, et al. Long-term treatment of acromegalic patients resistant to somatostatin analogues with the GH receptor antagonist pegvisomant: its efficacy in relation to gender and previous radiotherapy. *Eur J Endocrinol.* 2009;160(4):535–542.
24. van der Lely AJ, Biller BM, Brue T, Buchfelder M, Ghigo E, Gomez R, et al. Long-term safety of pegvisomant in patients with acromegaly: comprehensive review of 1288 subjects in ACROSTUDY. *J Clin Endocrinol Metab.* 2012;97(5):1589–1597.

25. Marazuela M, Paniagua AE, Gahete MD, Lucas T, Alvarez-Escola C, Manzanares R, et al. Somatotroph tumor progression during pegvisomant therapy: a clinical and molecular study. *J Clin Endocrinol Metab.* 2011;96(2):E251–E259.
26. Buchfelder M, Weigel D, Droste M, Mann K, Saller B, Brubach K, et al. Pituitary tumor size in acromegaly during pegvisomant treatment: experience from MR re-evaluations of the German Pegvisomant Observational Study. *Eur J Endocrinol.* 2009;161(1): 27–35.
27. Buhk JH, Jung S, Psychogios MN, Goricke S, Hartz S, Schulz-Heise S, et al. Tumor volume of growth hormone-secreting pituitary adenomas during treatment with pegvisomant: a prospective multicenter study. *J Clin Endocrinol Metab.* 2010;95(2):552–558.
28. Jimenez C, Burman P, Abs R, Clemmons DR, Drake WM, Hutson KR, et al. Follow-up of pituitary tumor volume in patients with acromegaly treated with pegvisomant in clinical trials. *Eur J Endocrinol.* 2008;159(5):517–523.
29. Freda PU, Gordon MB, Kelepouris N, Jonsson P, Koltowska-Haggstrom M, van der Lely AJ. Long-term treatment with pegvisomant as monotherapy in patients with acromegaly: experience from acrostudy. *Endocr Pract.* 2014;21:264–274.
30. Bernabeu I, Rodriguez-Gomez IA, Ramos-Levi AM, Marazuela M. Profile of pegvisomant in the management of acromegaly: an evidence based review of its place in therapy. *Res Rep Endocr Disord.* 2015;5:47–58.
31. Luque-Ramirez M, Paramo C, Varela da Costa C, Garcia-Mayor RV. Cost of management of invasive growth hormone-secreting macroadenoma. *J Endocrinol Invest.* 2007;30(7): 541–545.
32. Sowinski J, Sawicka N, Piatek K, Zybek A, Ruchala M. Pharmacoeconomic aspects of the treatment of pituitary gland tumours. *Contemp Oncol (Pozn).* 2013;17(2):137–143.
33. Moore DJ, Adi Y, Connock MJ, Bayliss S. Clinical effectiveness and cost-effectiveness of pegvisomant for the treatment of acromegaly: a systematic review and economic evaluation. *BMC Endocr Disord.* 2009;9:20.
34. Puig Domingo M. Treatment of acromegaly in the era of personalized and predictive medicine. *Clin Endocrinol (Oxf).* 2015;83:3–14.
35. Missale C, Nash SR, Robinson SW, Jaber M, Caron MG. Dopamine receptors: from structure to function. *Physiol Rev.* 1998;78(1):189–225.
36. Ferrari C, Barbieri C, Caldara R, Mucci M, Codecasa F, Paracchi A, et al. Long-lasting prolactin-lowering effect of cabergoline, a new dopamine agonist, in hyperprolactinemic patients. *J Clin Endocrinol Metab.* 1986;63(4):941–945.
37. Rickels MR, Snyder PJ. Cabergoline decreases somatotroph adenoma size: a case report. *Pituitary.* 2004;7(2):107–110.
38. Vilar L, Czepielewsk MA, Naves LA, Rollin GA, Casulari LA, Coelho CE. Substantial shrinkage of adenomas cosecreting growth hormone and prolactin with use of cabergoline therapy. *Endocr Pract.* 2007;13(4):396–402.
39. Ferrari C, Paracchi A, Romano C, Gerevini G, Boghen M, Barreca A, et al. Long-lasting lowering of serum growth hormone and prolactin levels by single and repetitive cabergoline administration in dopamine-responsive acromegalic patients. *Clin Endocrinol (Oxf).* 1988;29(5):467–476.
40. Bernabeu I, Alvarez-Escola C, Paniagua AE, Lucas T, Pavon I, Cabezas-Agricola JM, et al. Pegvisomant and cabergoline combination therapy in acromegaly. *Pituitary.* 2013;16(1):101–108.
41. Higham CE, Atkinson AB, Alywin S, Martin NM, Moyes VJ, Newell-Price J, *et al* A prospective clinical trial of combined cabergoline and pegvisomant treatment in patients with active acromegaly (Abstract). *Ninety-first Annual Meeting of the Endocrine Society*, June 10–13, 2009, Washington, DC; 2009
42. Verhelst JA, Abrams PJ, Abs R. Remission of acromegaly following long-term therapy with cabergoline: report of two cases. *Pituitary.* 2008;11(1):103–107.

43. Moyes VJ, Metcalfe KA, Drake WM. Clinical use of cabergoline as primary and adjunctive treatment for acromegaly. *Eur J Endocrinol.* 2008;159(5):541–545.
44. Sherlock M, Fernandez-Rodriguez E, Alonso AA, Reulen RC, Ayuk J, Clayton RN, et al. Medical therapy in patients with acromegaly: predictors of response and comparison of efficacy of dopamine agonists and somatostatin analogues. *J Clin Endocrinol Metab.* 2009;94(4):1255–1263.
45. Maiza JC, Vezzosi D, Matta M, Donadille F, Loubes-Lacroix F, Cournot M, et al. Long-term (up to 18 years) effects on GH/IGF-1 hypersecretion and tumour size of primary somatostatin analogue (SSTa) therapy in patients with GH-secreting pituitary adenoma responsive to SSTa. *Clin Endocrinol (Oxf).* 2007;67(2):282–289.
46. Mercado M, Borges F, Bouterfa H, Chang TC, Chervin A, Farrall AJ, et al. A prospective, multicentre study to investigate the efficacy, safety and tolerability of octreotide LAR (long-acting repeatable octreotide) in the primary therapy of patients with acromegaly. *Clin Endocrinol (Oxf).* 2007;66(6):859–868.
47. Freda PU, Katznelson L, van der Lely AJ, Reyes CM, Zhao S, Rabinowitz D. Long-acting somatostatin analog therapy of acromegaly: a meta-analysis. *J Clin Endocrinol Metab.* 2005;90(8):4465–4473.
48. Bevan JS. Clinical review: the antitumoral effects of somatostatin analog therapy in acromegaly. *J Clin Endocrinol Metab.* 2005;90(3):1856–1863.
49. Patel YC. Molecular pharmacology of somatostatin receptor subtypes. *J Endocrinol Invest.* 1997;20(6):348–367.
50. Patel YC. Somatostatin and its receptor family. *Front Neuroendocrinol.* 1999;20(3): 157–198.
51. Duran-Prado M, Saveanu A, Luque RM, Gahete MD, Gracia-Navarro F, Jaquet P, et al. A potential inhibitory role for the new truncated variant of somatostatin receptor 5, sst5TMD4, in pituitary adenomas poorly responsive to somatostatin analogs. *J Clin Endocrinol Metab.* 2010;95(5):2497–2502.
52. Cordoba-Chacon J, Gahete MD, Duran-Prado M, Luque RM, Castano JP. Truncated somatostatin receptors as new players in somatostatin-cortistatin pathophysiology. *Ann NY Acad Sci.* 2011;1220:6–15.
53. Casarini AP, Jallad RS, Pinto EM, Soares IC, Nonogaki S, Giannella-Neto D, et al. Acromegaly: correlation between expression of somatostatin receptor subtypes and response to octreotide-lar treatment. *Pituitary.* 2009;12(4):297–303.
54. Neto LV, Machado Ede O, Luque RM, Taboada GF, Marcondes JB, Chimelli LM, et al. Expression analysis of dopamine receptor subtypes in normal human pituitaries, non-functioning pituitary adenomas and somatotropinomas, and the association between dopamine and somatostatin receptors with clinical response to octreotide-LAR in acromegaly. *J Clin Endocrinol Metab.* 2009;94(6):1931–1937.
55. Zatelli MC. Antiproliferative effects of somatostatin analogs in endocrine tumours. *F1000 Med Rep.* 2009;8:1.pii: 40. doi: 10.3410/M1-40.
56. Taboada GF, Luque RM, Bastos W, Guimaraes RF, Marcondes JB, Chimelli LM, et al. Quantitative analysis of somatostatin receptor subtype (SSTR1-5) gene expression levels in somatotropinomas and non-functioning pituitary adenomas. *Eur J Endocrinol.* 2007;156(1):65–74.
57. Duran-Prado M, Gahete MD, Martinez-Fuentes AJ, Luque RM, Quintero A, Webb SM, et al. Identification and characterization of two novel truncated but functional isoforms of the somatostatin receptor subtype 5 differentially present in pituitary tumors. *J Clin Endocrinol Metab.* 2009;94(7):2634–2643.
58. Duran-Prado M, Gahete MD, Delgado-Niebla E, Martinez-Fuentes AJ, Vazquez-Martinez R, Garcia-Navarro S, et al. Truncated variants of pig somatostatin receptor subtype 5 (sst5) act as dominant-negative modulators for sst2-mediated signaling. *Am J Physiol Endocrinol Metab.* 2012;303(11):E1325–E1334.

59. Bhayana S, Booth GL, Asa SL, Kovacs K, Ezzat S. The implication of somatotroph adenoma phenotype to somatostatin analog responsiveness in acromegaly. *J Clin Endocrinol Metab.* 2005;90(11):6290–6295.
60. Fougner SL, Casar-Borota O, Heck A, Berg JP, Bollerslev J. Adenoma granulation pattern correlates with clinical variables and effect of somatostatin analogue treatment in a large series of patients with acromegaly. *Clin Endocrinol (Oxf).* 2012;76(1):96–102.
61. Kato M, Inoshita N, Sugiyama T, Tani Y, Shichiri M, Sano T, et al. Differential expression of genes related to drug responsiveness between sparsely and densely granulated somatotroph adenomas. *Endocr J.* 2012;59(3):221–228.
62. Chahal HS, Trivellin G, Leontiou CA, Alband N, Fowkes RC, Tahir A, et al. Somatostatin analogs modulate AIP in somatotroph adenomas: the role of the ZAC1 pathway. *J Clin Endocrinol Metab.* 2012;97(8):E1411–E1420.
63. Gadelha MR, Kasuki L, Korbonits M. Novel pathway for somatostatin analogs in patients with acromegaly. *Trends Endocrinol Metab.* 2013;24(5):238–246.
64. Kasuki L, Wildemberg LE, Neto LV, Marcondes J, Takiya CM, Gadelha MR. Ki-67 is a predictor of acromegaly control with octreotide LAR independent of SSTR2 status and relates to cytokeratin pattern. *Eur J Endocrinol.* 2013;169(2):217–223.
65. Ben-Shlomo A, Melmed S. Pituitary somatostatin receptor signaling. *Trends Endocrinol Metab.* 2010;21(3):123–133.
66. Lesche S, Lehmann D, Nagel F, Schmid HA, Schulz S. Differential effects of octreotide and pasireotide on somatostatin receptor internalization and trafficking *in vitro*. *J Clin Endocrinol Metab.* 2009;94(2):654–661.
67. Petersenn S, Bollerslev J, Arafat AM, Schopohl J, Serri O, Katznelson L, et al. Pharmacokinetics, pharmacodynamics, and safety of pasireotide LAR in patients with acromegaly: a randomized, multicenter, open-label, phase I study. *J Clin Pharmacol.* 2014;54(11):1308–1317.
68. Gadelha MR, Bronstein MD, Brue T, Coculescu M, Fleseriu M, Guitelman M, et al. Pasireotide versus continued treatment with octreotide or lanreotide in patients with inadequately controlled acromegaly (PAOLA): a randomised, phase 3 trial. *Lancet Diabetes Endocrinol.* 2014;2(11):875–884.
69. Colao A, Bronstein MD, Freda P, Gu F, Shen CC, Gadelha M, et al. Pasireotide versus octreotide in acromegaly: a head-to-head superiority study. *J Clin Endocrinol Metab.* 2014;99(3):791–799.
70. Goldenberg N, Barkan A. Factors regulating growth hormone secretion in humans. *Endocrinol Metab Clin North Am.* 2007;36(1):37–55.
71. Edens A, Talamantes F. Alternative processing of growth hormone receptor transcripts. *Endocr Rev.* 1998;19(5):559–582.
72. Barton DE, Foellmer BE, Wood WI, Francke U. Chromosome mapping of the growth hormone receptor gene in man and mouse. *Cytogenet Cell Genet.* 1989;50(2–3):137–141.
73. Pantel J, Machinis K, Sobrier ML, Duquesnoy P, Goossens M, Amselem S. Species-specific alternative splice mimicry at the growth hormone receptor locus revealed by the lineage of retroelements during primate evolution. *J Biol Chem.* 2000;275(25):18664–18669.
74. Gent J, van Kerkhof P, Roza M, Bu G, Strous GJ. Ligand-independent growth hormone receptor dimerization occurs in the endoplasmic reticulum and is required for ubiquitin system-dependent endocytosis. *Proc Natl Acad Sci USA.* 2002;99(15):9858–9863.
75. Brown RJ, Adams JJ, Pelekanos RA, Wan Y, McKinstry WJ, Palethorpe K, et al. Model for growth hormone receptor activation based on subunit rotation within a receptor dimer. *Nat Struct Mol Biol.* 2005;12(9):814–821.
76. Moutoussamy S, Kelly PA, Finidori J. Growth-hormone-receptor and cytokine-receptor-family signaling. *Eur J Biochem.* 1998;255(1):1–11.

77. Carter-Su C, Smit LS. Signaling via JAK tyrosine kinases: growth hormone receptor as a model system. *Recent Prog Horm Res.* 1998;53:61–82discussion 82–83.
78. Birzniece V, Sata A, Ho KK. Growth hormone receptor modulators. *Rev Endocr Metab Disord.* 2009;10(2):145–156.
79. Berg CA, Pokrajac A, Bidlingmaier M, Strasburger CJ, Shalet SM, Trainer PJ. Use of a GH receptor antagonist (GHRA) to explore the relationship between GH and IGF-I in adults with severe GH deficiency (GHD). *Clin Endocrinol (Oxf).* 2009;70(3):439–445.
80. Dos Santos C, Essioux L, Teinturier C, Tauber M, Goffin V, Bougneres P. A common polymorphism of the growth hormone receptor is associated with increased responsiveness to growth hormone. *Nat Genet.* 2004;36(7):720–724.
81. Urbanek M, Russell JE, Cooke NE, Liebhaber SA. Functional characterization of the alternatively spliced, placental human growth hormone receptor. *J Biol Chem.* 1993;268 (25):19025–19032.
82. Sobrier ML, Duquesnoy P, Duriez B, Amselem S, Goossens M. Expression and binding properties of two isoforms of the human growth hormone receptor. *FEBS Lett.* 1993;319(1–2):16–20.
83. Wassenaar MJ, Dekkers OM, Pereira AM, Wit JM, Smit JW, Biermasz NR, et al. Impact of the exon 3-deleted GH receptor polymorphism on baseline height and the growth response to recombinant human growth hormone therapy in growth hormone deficient (GHD) and non-GHD children with short stature: a systematic review and meta-analysis. *J Clin Endocrinol Metab.* 2009;94(10):3721–3730.
84. Meyer S, Schaefer S, Stolk L, Arp P, Uitterlinden AG, Plockinger U, et al. Association of the exon 3 deleted/full-length GHR polymorphism with recombinant growth hormone dose in growth hormone-deficient adults. *Pharmacogenomics.* 2009;10(10): 1599–1608.
85. van der Klaauw AA, van der Straaten T, Baak-Pablo R, Biermasz NR, Guchelaar HJ, Pereira AM, et al. Influence of the d3-growth hormone (GH) receptor isoform on short-term and long-term treatment response to GH replacement in GH-deficient adults. *J Clin Endocrinol Metab.* 2008;93(7):2828–2834.
86. Barbosa EJ, Palming J, Glad CA, Filipsson H, Koranyi J, Bengtsson BA, et al. Influence of the exon 3-deleted/full-length growth hormone (GH) receptor polymorphism on the response to GH replacement therapy in adults with severe GH deficiency. *J Clin Endocrinol Metab.* 2009;94(2):639–644.
87. Schmid C, Krayenbuehl PA, Bernays RL, Zwimpfer C, Maly FE, Wiesli P. Growth hormone (GH) receptor isoform in acromegaly: lower concentrations of GH but not insulin-like growth factor-1 in patients with a genomic deletion of exon 3 in the GH receptor gene. *Clin Chem.* 2007;53(8):1484–1488.
88. Kamenicky P, Dos Santos C, Espinosa C, Salenave S, Galland F, Le Bouc Y, et al. D3 growth hormone receptor polymorphism is not associated with IGF-1 levels in untreated acromegaly. *Eur J Endocrinol.* 2009;161(2):231–235.
89. Mercado M, Gonzalez B, Sandoval C, Esquenazi Y, Mier F, Vargas G, et al. Clinical and biochemical impact of the d3 growth hormone receptor genotype in acromegaly. *J Clin Endocrinol Metab.* 2008;93(9):3411–3415.
90. Bianchi A, Giustina A, Cimino V, Pola R, Angelini F, Pontecorvi A, et al. Influence of growth hormone receptor d3 and full-length isoforms on biochemical treatment outcomes in acromegaly. *J Clin Endocrinol Metab.* 2009;94(6):2015–2022.
91. Bianchi A, Mazziotti G, Tilaro L, Cimino V, Veltri F, Gaetani E, et al. Growth hormone receptor polymorphism and the effects of pegvisomant in acromegaly. *Pituitary.* 2009;12 (3):196–199.
92. Wassenaar MJ, Biermasz NR, Pereira AM, van der Klaauw AA, Smit JW, Roelfsema F, et al. The exon-3 deleted growth hormone receptor polymorphism predisposes to long-term complications of acromegaly. *J Clin Endocrinol Metab.* 2009;94(12):4671–4678.

93. Montefusco L, Filopanti M, Ronchi CL, Olgiati L, La-Porta C, Losa M, et al. d3-Growth hormone receptor polymorphism in acromegaly: effects on metabolic phenotype. *Clin Endocrinol (Oxf)*. 2010;72(5):661–667.
94. Bernabeu I, Alvarez-Escola C, Quinteiro C, Lucas T, Puig-Domingo M, Luque-Ramirez M, et al. The exon 3-deleted growth hormone receptor is associated with better response to pegvisomant therapy in acromegaly. *J Clin Endocrinol Metab*. 2010;95 (1):222–229.
95. Filopanti M, Olgiati L, Mantovani G, Corbetta S, Arosio M, Gasco V, et al. Growth hormone receptor variants and response to pegvisomant in monotherapy or in combination with somatostatin analogs in acromegalic patients: a multicenter study. *J Clin Endocrinol Metab*. 2012;97(2):E165–E172.
96. Cinar N, Dagdelen S, Yorgun H, Canpolat U, Kabakci G, Erbas T. The clinical and cardiometabolic effects of d3-growth hormone receptor polymorphism in acromegaly. *Pituitary*. 2014;18:116–125.
97. Mormando M, Nasto LA, Bianchi A, Mazziotti G, Giampietro A, Pola E, et al. GH receptor isoforms and skeletal fragility in acromegaly. *Eur J Endocrinol*. 2014;171(2): 237–245.
98. Filopanti M, Giavoli C, Grottoli S, Bianchi A, De Marinis L, Ghigo E, et al. The exon 3-deleted growth hormone receptor: molecular and functional characterization and impact on GH/IGF-I axis in physiological and pathological conditions. *J Endocrinol Invest*. 2011;34(11):861–868.
99. Bougneres P. The exon-3 deletion of the growth hormone receptor (GHR) gene still has a limited impact in clinical endocrinology. *J Clin Endocrinol Metab*. 2010;95(1):56–59.
100. Asa SL, Coschigano KT, Bellush L, Kopchick JJ, Ezzat S. Evidence for growth hormone (GH) autoregulation in pituitary somatotrophs in GH antagonist-transgenic mice and GH receptor-deficient mice. *Am J Pathol*. 2000;156(3):1009–1015.
101. Lichanska AM, Waters MJ. New insights into growth hormone receptor function and clinical implications. *Horm Res*. 2008;69(3):138–145.
102. Bass SH, Mulkerrin MG, Wells JA. A systematic mutational analysis of hormone-binding determinants in the human growth hormone receptor. *Proc Natl Acad Sci USA*. 1991;88(10):4498–4502.
103. Asa SL, Digiovanni R, Jiang J, Ward ML, Loesch K, Yamada S, et al. A growth hormone receptor mutation impairs growth hormone autofeedback signaling in pituitary tumors. *Cancer Res*. 2007;67(15):7505–7511.
104. Kola B, Korbonits M, Diaz-Cano S, Kaltsas G, Morris DG, Jordan S, et al. Reduced expression of the growth hormone and type 1 insulin-like growth factor receptors in human somatotroph tumours and an analysis of possible mutations of the growth hormone receptor. *Clin Endocrinol (Oxf)*. 2003;59(3):328–338.
105. Frank SJ, Fuchs SY. Modulation of growth hormone receptor abundance and function: roles for the ubiquitin-proteasome system. *Biochim Biophys Acta*. 2008;1782(12): 785–794.
106. Mete O, Ezzat S, Asa SL. Biomarkers of aggressive pituitary adenomas. *J Mol Endocrinol*. 2012;49(2):R69–R78.
107. van der Lely AJ, Biller BMK, Brue T, Buchfelder M, Ghigo E, Gomez R, et al. Long-term safety of pegvisomant in patients with acromegaly: comprehensive review of 1288 subjects in ACROSTUDY. *J Clin Endocrinol Metab* 2012;97:1589–1597.
108. Trainer PJ. ACROSTUDY: the first 5 years. *Eur J Endocrinol*. 2009;161(suppl 1): S19–S24.
109. Dastot F, Sobrier ML, Duquesnoy P, Duriez B, Goossens M, Amselem S. Alternatively spliced forms in the cytoplasmic domain of the human growth hormone (GH) receptor regulate its ability to generate a soluble GH-binding protein. *Proc Natl Acad Sci USA*. 1996;93(20):10723–10728.

110. Ballesteros M, Leung KC, Ross RJ, Iismaa TP, Ho KK. Distribution and abundance of messenger ribonucleic acid for growth hormone receptor isoforms in human tissues. *J Clin Endocrinol Metab.* 2000;85(8):2865–2871.
111. Ross RJ, Esposito N, Shen XY, Von Laue S, Chew SL, Dobson PR, et al. A short isoform of the human growth hormone receptor functions as a dominant negative inhibitor of the full-length receptor and generates large amounts of binding protein. *Mol Endocrinol.* 1997;11(3):265–273.
112. Rowland JE, Lichanska AM, Kerr LM, White M, d'Aniello EM, Maher SL, et al. *In vivo* analysis of growth hormone receptor signaling domains and their associated transcripts. *Mol Cell Biol.* 2005;25(1):66–77.
113. Amit T, Youdim MB, Hochberg Z. Clinical review 112: does serum growth hormone (GH) binding protein reflect human GH receptor function? *J Clin Endocrinol Metab.* 2000;85(3):927–932.
114. Amit T, Bergman T, Dastot F, Youdim MB, Amselem S, Hochberg Z. A membrane-fixed, truncated isoform of the human growth hormone receptor. *J Clin Endocrinol Metab.* 1997;82(11):3813–3817.
115. Fisker S. Physiology and pathophysiology of growth hormone-binding protein: methodological and clinical aspects. *Growth Horm IGF Res.* 2006;16(1):1–28.
116. Sotiropoulos A, Goujon L, Simonin G, Kelly PA, Postel-Vinay MC, Finidori J. Evidence for generation of the growth hormone-binding protein through proteolysis of the growth hormone membrane receptor. *Endocrinology.* 1993;132(4):1863–1865.
117. Leung DW, Spencer SA, Cachianes G, Hammonds RG, Collins C, Henzel WJ, et al. Growth hormone receptor and serum binding protein: purification, cloning and expression. *Nature.* 1987;330(6148):537–543.
118. Zhang Y, Jiang J, Black RA, Baumann G, Frank SJ. Tumor necrosis factor-alpha converting enzyme (TACE) is a growth hormone binding protein (GHBP) sheddase: the metalloprotease TACE/ADAM-17 is critical for (PMA-induced) GH receptor proteolysis and GHBP generation. *Endocrinology.* 2000;141(12):4342–4348.
119. Mullis PE, Wagner JK, Eble A, Nuoffer JM, Postel-Vinay MC. Regulation of human growth hormone receptor gene transcription by human growth hormone binding protein. *Mol Cell Endocrinol.* 1997;131(1):89–96.
120. Mullis PE, Eble A, Wagner JK, Holl RW, Silbergeld A, Laron Z. Effect of different serum concentrations of growth hormone-binding protein (GHBP) on the regulation of GH receptor/GHBP gene transcription in a human hepatoma cell line. *Horm Res.* 1997;47(2):73–80.
121. Leung KC, Johannsson G, Leong GM, Ho KK. Estrogen regulation of growth hormone action. *Endocr Rev.* 2004;25(5):693–721.
122. Harding PA, Wang X, Okada S, Chen WY, Wan W, Kopchick JJ. Growth hormone (GH) and a GH antagonist promote GH receptor dimerization and internalization. *J Biol Chem.* 1996;271(12):6708–6712.
123. Maamra M, Kopchick JJ, Strasburger CJ, Ross RJ. Pegvisomant, a growth hormone-specific antagonist, undergoes cellular internalization. *J Clin Endocrinol Metab.* 2004;89(9):4532–4537.
124. Gent J, van Kerkhof P, Roza M, Bu G, Strous GJ. Ligand-independent growth hormone receptor dimerization occurs in the endoplasmic reticulum and is required for ubiquitin system-dependent endocytosis. *Proc Natl Acad Sci USA.* 2002;99(15):9858–9863.
125. Higham CE, Trainer PJ. Growth hormone excess and the development of growth hormone receptor antagonists. *Exp Physiol.* 2008;93(11):1157–1169.
126. Ross RJ, Leung KC, Maamra M, Bennett W, Doyle N, Waters MJ, et al. Binding and functional studies with the growth hormone receptor antagonist, B2036-PEG (pegvisomant), reveal effects of pegylation and evidence that it binds to a receptor dimer. *J Clin Endocrinol Metab.* 2001;86(4):1716–1723.

127. Rosen CJ, Kurland ES, Vereault D, Adler RA, Rackoff PJ, Craig WY, et al. Association between serum insulin growth factor-I (IGF-I) and a simple sequence repeat in IGF-I gene: implications for genetic studies of bone mineral density. *J Clin Endocrinol Metab.* 1998;83(7):2286–2290.
128. Vaessen N, Heutink P, Janssen JA, Witteman JC, Testers L, Hofman A, et al. A polymorphism in the gene for IGF-I: functional properties and risk for type 2 diabetes and myocardial infarction. *Diabetes.* 2001;50(3):637–642.
129. Vaessen N, Janssen JA, Heutink P, Hofman A, Lamberts SW, Oostra BA, et al. Association between genetic variation in the gene for insulin-like growth factor-I and low birthweight. *Lancet.* 2002;359(9311):1036–1037.
130. Rietveld I, Janssen JA, Hofman A, Pols HA, van Duijn CM, Lamberts SW. A polymorphism in the IGF-I gene influences the age-related decline in circulating total IGF-I levels. *Eur J Endocrinol.* 2003;148(2):171–175.
131. Rietveld I, Janssen JA, van Rossum EF, Houwing-Duistermaat JJ, Rivadeneira F, Hofman A, et al. A polymorphic CA repeat in the IGF-I gene is associated with gender-specific differences in body height, but has no effect on the secular trend in body height. *Clin Endocrinol (Oxf).* 2004;61(2):195–203.
132. Bleumink GS, Rietveld I, Janssen JA, van Rossum EF, Deckers JW, Hofman A, et al. Insulin-like growth factor-I gene polymorphism and risk of heart failure (the Rotterdam Study). *Am J Cardiol.* 2004;94(3):384–386.
133. Costalonga EF, Antonini SR, Guerra Jr G, Coletta RR, Franca MM, Braz AF, et al. Growth hormone pharmacogenetics: the interactive effect of a microsatellite in the IGF1 promoter region with the GHR-exon 3 and -202 A/C IGFBP3 variants on treatment outcomes of children with severe GH deficiency. *Pharmacogenomics J.* 2012;12(5):439–445.
134. Miletta MC, Scheidegger UA, Giordano M, Bozzola M, Pagani S, Bona G, et al. Association of the (CA)n repeat polymorphism of insulin-like growth factor-I and -202 A/C IGF-binding protein-3 promoter polymorphism with adult height in patients with severe growth hormone deficiency. *Clin Endocrinol (Oxf).* 2012;76(5):683–690.
135. Akin F, Turgut S, Cirak B, Kursunluoglu R. IGF(CA)19 and IGFBP-3-202A/C gene polymorphism in patients with acromegaly. *Growth Horm IGF Res.* 2010;20(6): 399–403.
136. Ramos-Levi AM, Marazuela M, Paniagua A, Quinteiro C, Riveiro J, Alvarez-Escola C, et al. Analysis of IGF(CA)19 and IGFBP3-202A/C gene polymorphisms in patients with acromegaly: association with clinical presentation and response to treatments. *Eur J Endocrinol.* 2015;172(2):115–122.
137. Deal C, Ma J, Wilkin F, Paquette J, Rozen F, Ge B, et al. Novel promoter polymorphism in insulin-like growth factor-binding protein-3: correlation with serum levels and interaction with known regulators. *J Clin Endocrinol Metab.* 2001;86(3): 1274–1280.
138. Harrela M, Koistinen H, Kaprio J, Lehtovirta M, Tuomilehto J, Eriksson J, et al. Genetic and environmental components of interindividual variation in circulating levels of IGF-I, IGF-II, IGFBP-1, and IGFBP-3. *J Clin Invest.* 1996;98(11):2612–2615.
139. van der Kaay DC, Hendriks AE, Ester WA, Leunissen RW, Willemsen RH, de Kort SW, et al. Genetic and epigenetic variability in the gene for IGFBP-3 (IGFBP3): correlation with serum IGFBP-3 levels and growth in short children born small for gestational age. *Growth Horm IGF Res.* 2009;19(3):198–205.
140. Costalonga EF, Antonini SR, Guerra-Junior G, Mendonca BB, Arnhold IJ, Jorge AA. The -202 A allele of insulin-like growth factor binding protein-3 (IGFBP3) promoter polymorphism is associated with higher IGFBP-3 serum levels and better growth response to growth hormone treatment in patients with severe growth hormone deficiency. *J Clin Endocrinol Metab.* 2009;94(2):588–595.

141. Braz AF, Costalonga EF, Montenegro LR, Trarbach EB, Antonini SR, Malaquias AC, et al. The interactive effect of GHR-exon 3 and -202 A/C IGFBP3 polymorphisms on rhGH responsiveness and treatment outcomes in patients with Turner syndrome. *J Clin Endocrinol Metab.* 2012;97(4):E671–E677.
142. Hodish I, Barkan A. Long-term effects of pegvisomant in patients with acromegaly. *Nat Clin Pract Endocrinol Metab.* 2008;4(6):324–332.
143. Biering H, Saller B, Bauditz J, Pirlich M, Rudolph B, Johne A, et al. Elevated transaminases during medical treatment of acromegaly: a review of the German pegvisomant surveillance experience and a report of a patient with histologically proven chronic mild active hepatitis. *Eur J Endocrinol.* 2006;154(2):213–220.
144. Feenstra J, van Aken MO, de Herder WW, Feelders RA, van der Lely AJ. Drug-induced hepatitis in an acromegalic patient during combined treatment with pegvisomant and octreotide long-acting repeatable attributed to the use of pegvisomant. *Eur J Endocrinol.* 2006;154(6):805–806.
145. Bernabeu I, Cameselle-Teijeiro J, Casanueva F, Marazuela M. Pegvisomant-induced cholestatic hepatitis with jaundice in a patient with Gilbert's syndrome. *Eur J Endocrinol.* 2009;160(5):869–872.
146. Soto Moreno A, Guerrero Vazquez R, Venegas Moreno E, Palma Milla S, Castano JP, Leal Cerro A. Self-limited acute hepatotoxicity caused by pegvisomant. *Pituitary.* 2009;14:371–376.
147. Neggers SJ, de Herder WW, Janssen JA, Feelders RA, van der Lely AJ. Combined treatment for acromegaly with long-acting somatostatin analogs and pegvisomant: long-term safety for up to 4.5 years (median 2.2 years) of follow-up in 86 patients. *Eur J Endocrinol.* 2009;160(4):529–533.
148. Verma S, Kaplowitz N. Diagnosis, management and prevention of drug-induced liver injury. *Gut.* 2009;58(11):1555–1564.
149. Kaplowitz N. Drug-induced liver injury. *Clin Infect Dis.* 2004;38(suppl 2):S44–S48.
150. Strassburg CP. Pharmacogenetics of Gilbert's syndrome. *Pharmacogenomics.* 2008;9(6): 703–715.
151. Strassburg CP, Lankisch TO, Manns MP, Ehmer U. Family 1 uridine-5′-diphosphate glucuronosyltransferases (UGT1A): from Gilbert's syndrome to genetic organization and variability. *Arch Toxicol.* 2008;82(7):415–433.
152. Bosma PJ, Chowdhury JR, Bakker C, Gantla S, de Boer A, Oostra BA, et al. The genetic basis of the reduced expression of bilirubin UDP-glucuronosyltransferase 1 in Gilbert's syndrome. *N Engl J Med.* 1995;333(18):1171–1175.
153. Burchell B, Hume R. Molecular genetic basis of Gilbert's syndrome. *J Gastroenterol Hepatol.* 1999;14(10):960–966.
154. Burchell B, Soars M, Monaghan G, Cassidy A, Smith D, Ethell B. Drug-mediated toxicity caused by genetic deficiency of UDP-glucuronosyltransferases. *Toxicol Lett.* 2000;112–113:333–340.
155. FDA. Pegvisomant FDA Approval Package. Clinical Pharmacology and Biopharmaceutics review and Medical review. http://wwwaccessdatafdagov/drugsatfda_docs/nda/2003/021106_somavertcfm; 2014 (Accessed Nov. 2014)
156. Bernabeu I, Marazuela M, Lucas T, Loidi L, Alvarez-Escola C, Luque-Ramirez M, et al. Pegvisomant-induced liver injury is related to the UGT1A*28 polymorphism of Gilbert's syndrome. *J Clin Endocrinol Metab.* 2010;95(5):2147–2154.
157. Filopanti M, Barbieri AM, Mantovani G, Corbetta S, Gasco V, Ragonese M, et al. Role of uridine diphosphate-5′-glucuronosyltransferase type 1A (UGT1A1) and alcohol dehydrogenase (ADH) gene polymorphisms in pegvisomant-induced liver toxicity in acromegalic patients. *Eur J Endocrinol.* 2013;170:247–254.

CHAPTER SIX

Neurosurgical Treatment of Acromegaly

Michael Buchfelder[1], Julian Feulner
Department of Neurosurgery, University of Erlangen-Nürnberg, Erlangen, Germany
[1] Corresponding author: e-mail: michael.buchfelder@uk-erlangen.de.

Contents

Abstract

Surgical removal of as much tumor mass as possible is usually considered the first step of treatment in acromegaly, unless the patients are unfit for surgery or refuse an operation. To date, in almost all cases, minimally invasive, transsphenoidal microscopic or endoscopic approaches are used. Whether a curative approach is feasible or a debulking procedure is planned, can be anticipated on the basis of preoperative magnetic resonance imaging. It mostly depends on localization, size, and the invasive character of the lesion. The surgical results depend on tumor-related factors such as size, extension, the presence or absence of invasion, and the magnitude of IGF-1 and growth hormone oversecretion, respectively. However, even surgeon-related factors such as experience and case load of the centers have been shown to strongly affect surgical results and complication rates. A reoperation can be considered at various stages in the treatment algorithm. There are several new technical gadgets which might aid in the surgical procedure: navigation, the Doppler probe, and variants of intraoperative imaging.

Progress in Molecular Biology and Translational Science, Volume 138
ISSN 1877-1173
http://dx.doi.org/10.1016/bs.pmbts.2015.11.002

1. INTRODUCTION

Acromegaly is a rare but severe disease with potential to threaten a patient's life. Mortality is significantly higher in acromegaly compared to the general population, as several retrospective cohort studies have revealed.[1] Death is mostly caused by cardiac and vascular reasons. However, mortality is impressively decreased by a satisfactory reduction in GH levels, as documented suggested by several studies.[2–5] After confirmation of the diagnosis it is consequently necessary to lower serum GH levels and IGF-1 levels, respectively, by an aggressive management. Surgery remains the primary therapy for most patients. To date it is standardized, safe, and efficient after optimization over a 100 years.[6,7] Furthermore, surgery is a comparatively cheap and often permanent treatment although additional therapies or irradiation are required postoperatively for many patients.[8–10]

Normalization of GH and IGF-1 secretion levels while simultaneously preserving the normal pituitary function as well as eliminating the tumor's mass effect are the goals of surgical therapy in acromegaly. Normalizing mortality with reduction of the hormonal secretion is ideally achieved while avoiding complications. Over the years the definition of disease remission in acromegaly has changed dramatically. GH assays were improved and became more sensitive. Modern definitions of cure were defined by International Consensus Conferences. The remission criteria defined by the Cortina d́ Ampezzo Consensus Conference were used in most of the recently published surgical series[11] with minor adaptations suggested by the US Endocrine Society's most recent guidelines for the management of acromegaly.[12] Remission was thus defined by a random serum GH level below 2.5 μg/L accompanied by a suppression of GH to below 1 ng/mL following oral glucose load, and a normal age-adjusted IGF-1. They were based on hormone determinations with commercially available standardized assays. Routinely, endocrinological evaluation which decides upon remission or persistent disease is performed only several weeks after surgery. In current Guidelines[12] more stringent remission criteria (undetectable GH level, normal IGF-1) are recommended.

2. INDICATIONS FOR SURGERY

The most recent Guidelines established by the Endocrine Society[12] claim stringent criteria for the diagnosis of acromegaly. Random GH levels are not recommended to use for diagnosing acromegaly. Instead, measuring of IGF-1 and—if IGF-1 serum levels are elevated or equivocal—confirmation of diagnosis by finding lack of suppression of GH levels < 1 μg/L following oral glucose load are suggested.

In the past and even more recently, surgery was considered the primary treatment of acromegaly in most patients.[12–17] This sequence has been maintained even in the newly published algorithms.[10,12] However, with potent drugs now being available which inhibit GH-secretion and/or lower IGF-1, there are alternatives for selected patients with high expected perioperative morbidity or for those with large and invasive tumors in which only a minor portion is considered resectable.[8,9] Generally, tumors with enclosed nature and moderate size are expected to be totally resectable by expert surgeons and remain excellent candidates for primary surgical therapy.[18–20] Moreover, debulking of large and invasive tumors prior to medical treatment or irradiation improves the efficacy of subsequent medical therapy,[21–24] at least as an immediate effect. One would suppose that the long-term results are also improved by debulking surgery. Thus, the indication for surgery in acromegaly is generally established, once the diagnosis of acromegaly is confirmed, provided that the general condition and the comorbidities of the patient allow for anesthesia and surgery.[10,12,25]

3. PREOPERATIVE EVALUATION

Endocrinological evaluation is required to provide information about the magnitude of GH and IGF-1 hypersecretion and potential hypopituitarism. The patient's history should provide information about previous operations of the nose or the sphenoid sinus. Generally, magnetic resonance imaging (MRI) is the optimal method to depict the tumor, because of its high soft tissue contrast resolution. For the evaluation of the tumor localization T1-weighted images in sagittal and coronal planes are required, before and after contrast enhancement. Additional information may be provided by other MRI sequences.[26] MR-angiography illustrates the localization and course of the major brain-supplying intracranial arteries. Thin-collimation CT scans with axial and/or coronal image reconstructions can provide additional and useful anatomical orientation in specific cases, since they reliably show osseous structures, like septations of the paranasal sinuses, more precisely. Plain skull X-rays are considered obsolete by most of the pituitary surgeons. In any tumor with development in the suprasellar space, vision should be ophthalmologically assessed by perimetry and testing of visual acuity.[6,7,25] Especially in the case of acromegaly any patient needs to be assessed for associated comorbidities[27] which may potentially influence the individual perioperative risk, such as arterial hypertension, glucose intolerance/diabetes, cardiac and respiratory disease. Additionally, a careful anesthesiological evaluation for possible airway management problems (i.e., due to macroglossia) is highly recommended.[28]

4. SURGICAL TECHNIQUES

To date, almost all pituitary tumor operations in patients with acromegaly are performed via the transsphenoidal route, which is a highly versatile approach. More than 95% of pituitary adenomas can be treated by using this approach.[25] Multiple variations are possible, starting with the patient's positioning. While some surgeons, like ourselves, prefer to operate on a patient in supine position with the head slightly extended (Fig. 1), a semisitting position[6,7] is favored by others (Fig. 2). Radiofluoroscopic

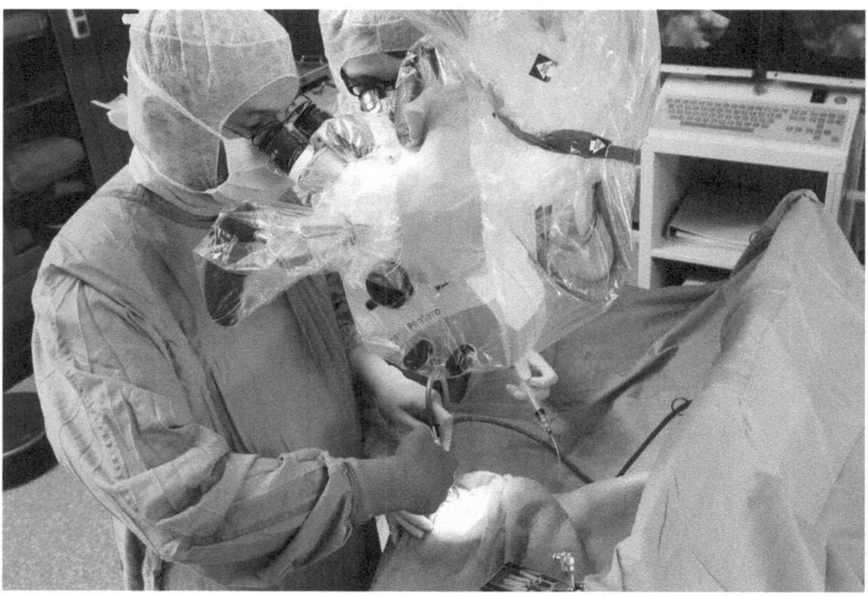

Figure 1 Transsphenoidal surgery to date: the patient is in supine position using a surgical high-resolution microscope and an image intensifier (c-arm) for intraoperative imaging and navigation.

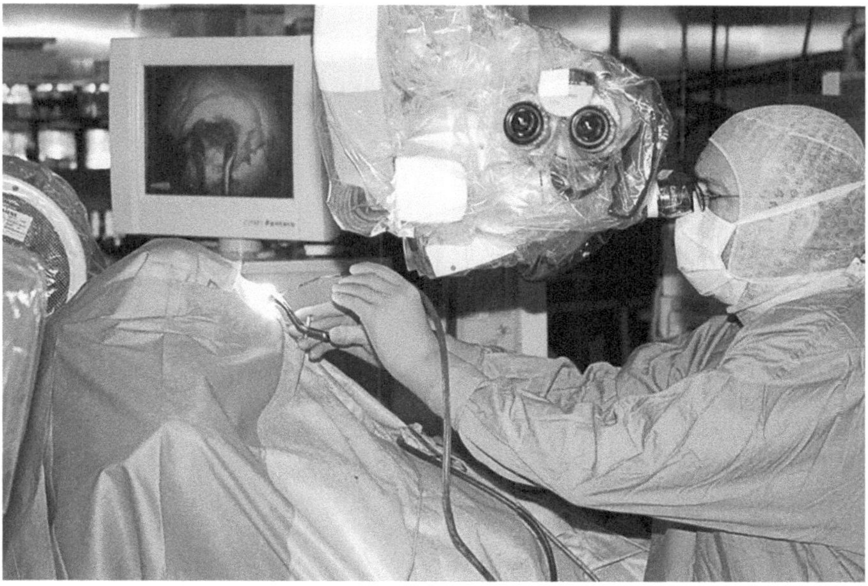

Figure 2 An alternative for positioning the patient for transsphenoidal surgery: a semisitting position.

control using an X-ray image intensifier is still the most frequently used procedure for both, navigation and intraoperative imaging. However, techniques using neuronavigation are used more and more frequently and provide more specific information. Transsphenoidal surgery can be performed with and without dissection of the nasal septal mucosa. Dissection of a submucosal tunnel, like commonly performed in nasal septum surgeries, is usually performed following a medial nasal incision. In patients with acromegaly only very rarely a sublabial incision is required. After the incision the mucosa is unilaterally separated from the cartilaginous and osseous nasal septae. The mucosal tunnel is kept open by a nasal speculum. At this point, usually the operating microscope is brought into place. For midline orientation the vomer can be used as a landmark. Subsequently the mucosa is coagulated and further separated from osseous structures laterally from the midline. After exposition of the anterior wall of the sphenoid sinus a sphenoidotomy is now being performed. The whole mucosa within the sphenoid sinus and septations are resected. At this stage the sellar floor can usually be visualized through the sphenoid sinus (transsphenoidally). In the case of incomplete pneumatization of the sphenoid sinus, extensive drilling is needed to expose the whole sellar floor to be able to reach the tumor's posterior basal surface. The sellar floor is then being resected and, for reasons of hemostasis and disinfection, hydrogen peroxide is temporarily introduced into the sphenoid sinus. The exposed basal dura of the sella is incised. Often soft tumors immediately prelapses through the dural opening. For further resection, variously shaped curettes and microforceps are available. The normal pituitary gland, which is often deformed by compression against the posterior portion of the fossa or the cavernous sinus by the tumor's mass effect can be identified by a yellowish color with fine vascular structures on its surface as well as a more solid consistency. Every possible attempt must be made to preserve the normal pituitary function. The extent of tumor resection is estimated by inspection and palpation of the tumor cavity. After resection often the arachnoid which covers the superior tumor surface decends into the intrasellar space. Its smooth surface can easily be recognized. Nevertheless, whenever it descends in several folds, each fold has to be addressed separately since residual tumor may be hidden within the pouches. It is easier to resect a large adenoma, if the connection between intra- and extrasellar tumor portions is wide than if it is narrow. Invasive nature of a tumor restricts its resectability. Local invasion may be successfully addressed by using adequate surgical techniques. In the case of invasion of the mucosa of the sphenoid sinus by the tumor, it can be easily radically resected. In the case of invasion of the skull base by the tumor, the bone can be drilled. Parasellar tumor proportions can be traced along the perforations of the medial wall of the cavernous sinus that have been created by the invading tumor. Tumor located laterally from the carotid artery can usually not be resected, and thus will remain. In the case of small microadenomas, section of the normal pituitary gland will be necessary in order to find the tumor, just like in adenomas associated with Cushing's disease.[6] The term "enlarged adenomectomy" refers to a small layer of normal pituitary gland (called "periadenoma"), which is additionally resected around the adenoma. The best overview of a tumor cavity after performed selective adenomectomy can usually be achieved in case of microadenomas and smaller macroadenomas. Even with utmost experience and optimal technical equipment there are still pituitary adenomas which cannot be resected completely. Mirrors and particularly endoscopes allow a better visualization of tumor portions and anatomical structures, especially those structures localized lateral to the midline. Reconstruction of the sellar floor can be performed using several methods. As long as there was no CSF leakage during the surgery in a small or moderately sized tumor, gelfoam may be enough. In case the arachnoid fold is very thin or CSF leaks have occurred intraoperatively, the sellar floor should be reconstructed with an autologous graft (fat, fascia lata, muscle, cartilage, or bone). To attach the graft it may be helpful to use human fibrin glow or other adhesives. The use of implants has been proposed, but is not employed by the authors. A lot of authors use a regimen of external CSF drainage for a few days after surgery in case of intraoperative CSF leakage. In microsurgical operations with dissection of the nasal septal mucosa and usage of a speculum, normally nasal tamponades are required.

As an alternative, a direct pernasal approach to the sphenoid sinus can be chosen. For this approach the speculum is guided under radiofluorescopic control or neuronavigation to the rostrum of the vomer. In this case the required mucosal incision is performed just in front of the vomer. The osseous nasal septum is then being fractured and the operation is continued as described earlier.[6] This direct pernasal approach is used by endoscopic transsphenoidal surgeons. They do not need a nasal speculum. If comparing endoscopic versus microsurgical techniques there are advantages and disadvantages to be pondered.[29,30] A recent study comparing the microscopic and endoscopic technique in acromegaly, each one carried out by an expert surgeon, revealed no difference in the surgical results[31] based on the currently available data. The Endocrine Society Guidelines state that there is no superiority of one of the techniques.[12] After all, whenever the use of the endoscope is considered useful to enhance visualization, it can be used during a microsurgical procedure.[25] Still, most data in the results section largely derive from microsurgery series. Not a long time ago, retrosellar tumor extension or development to the planum sphenoidale was considered as a contraindication for transsphenoidal surgery. With performing the extended transsphenoidal approach, this is no longer the case. "Extended" may involve opening to the tuberculum sellae or drilling through the clivus. Thus, large tumors, which were not accessible by transsphenoidal surgery in the past, can now be addressed via the nasal route as well. Unfortunately, one of the disadvantages of the described extended transsphenoidal approach is a high occurrence of postoperative CSF leaks.[29]

Less than 5% of patients, namely those who harbor tumors with extreme asymmetrical intracranial extension and poor communication with the intrasellar portions, still require a tumor resection by using transcranial approaches to reduce the tumor mass.[7,20,25,32,33] Usually, frontal or frontotemporal craniotomies are used below curvilinear incisions behind the hairline. The bone-flap is raised and, after dural opening, the dissection of the tumor is performed below the frontal and temporal lobes, respectively. By drilling of the sphenoid wing, aspiration of CSF, and the use of self-retaining retractors the required space for dissection is created. The carotid artery with its branches and visual pathways is exposed. Consequently an incision and central decompression of the tumor is performed, followed by the dissection of the tumor tissue from the arachnoidal layers of the basal cisterns, cranial nerves, blood vessels, and the infundibulum under direct vision through the microscope. Resecting a tumor completely during transcranial surgery while preserving the normal pituitary gland and its function is much more difficult than in transsphenoidal surgery.[25,32]

5. SURGICAL OUTCOME

Resulting from changing definitions of postoperative endocrinological remission,[2,3,11,15,17,27,34–37] success rates for "hormonal normalization" differ significantly over time in the published series from different institutions (Table 1).

For early series, the underlying criterion for disease remission was a postoperative random GH level below 5 μg/L. Ross *et al.*[15] and Zervas *et al.*[17] some 30 years ago reported on "normalization" rates for acromegaly between 56 and 66% after transsphenoidal surgery. Certainly, nobody would use such obsolete remission criteria anymore since it was recognized that very low random levels during the day are considered normal. Using the GH response after an oral glucose load to assess the efficiency of surgery, rates of endocrinological remission were found lower, as in a previous series from our institution published in 1992.[32] The remission rate of 71% reported by this study with a basal GH level below 5 μg/L as the criterion, dropped significantly to 57% when a glucose-suppressed GH level below 2 μg/L was used for the definition of endocrinological remission. Other series, such as those published by Losa *et al.*[38] and Valdemarsson *et al.*[47,48] suggested similar results. Tindall *et al.*,[16] Laws *et al.*,[19] Kreutzer *et al.*,[18] and, more recently, Nomikos *et al.*[20] showed that using all three criteria, including the normalization of basal IGF-1, does not change the reported endocrinological remission rates very much. In most patients a suppression of GH levels to below 1 ng/mL is associated with age- and sex-adjusted normal IGF-1 levels. A few individual patients still show elevated levels of IGF-1 even if GH levels could be suppressed to below 1 μg/L during OGTT and vice versa. Furthermore, the role of postoperative IGF-1 levels for the long-term outcome of acromegaly has not yet been fully elucidated, since assays used to determine IGF-1 levels changed over time. In assessing the outcome of

Table 1 Results of Primary Transsphenoidal Surgery for GH-Secreting Pituitary Adenomas.

Series	No. of Cases	Total Cure Rate (%)	Microadenomas	Macroadenomas	Criteria for Remission
Ross and Wilson (1988)[15]	153	56	N/A	N/A	GH < 5 μg/L
Losa et al. (1989)[38]	29	55	N/A	N/A	GH < 1 μg/L or GH < 1 μg/L (OGTT), normal IGF-1 level
Fahlbusch et al. (1992)[32]	222	57	72	49	GH < 2 μg/L (OGTT)
		71	81	65	GH < 5 μg/L
Tindall et al. (1993)[16]	91	82	N/A	N/A	GH < 5 μg/L and/or normal IGF-1 level
Davis et al. (1993)[39]	174	52	N/A	N/A	GH < 2 μg/L (basal or OGTT)
Sheaves et al. (1996)[40]	100	42	61	23	GH < 2.5 μg/L
Abosch et al. (1998)[34]	254	76	75	71	GH < 5 μg/L
Freda et al. (1998)[41]	115	61	88	53	GH < 2 μg/L (OGTT) or normal IGF-1 level
Laws et al. (2000)[19]	117	67	87	51	GH < 2.5 μg/L or GH < 1 μg/L (OGTT), normal IGF-1 level
Kreutzer et al. (2001)[18]	57	70	N/A	N/A	GH < 2.5 μg/L or GH < 1 μg/L (OGTT), normal IGF-1 level
De et al. (2003)[37]	90	54	72	50	GH < 2.5 μg/L or GH < 1 μg/L (OGTT), normal IGF-1 level
Mortini et al. (2005)[33]	320	59	83	53	GH < 1 μg/L (OGTT), normal IGF-1 level

(*Continued*)

Table 1 Results of Primary Transsphenoidal Surgery for GH-Secreting Pituitary Adenomas.—cont'd.

Series	No. of Cases	Total Cure Rate (%)	Microadenomas	Macroadenomas	Criteria for Remission
Nomikos et al. (2005)[20]	490	56	78	50	GH ≤ 2.5 μg/L, GH ≤ 1 μg/L (OGTT), normal IGF-1 level
Frank et al. (2006)[42]	16	44	N/A	N/A	GH ≤ 2.5 μg/L, GH ≤ 1 μg/L (OGTT), normal IGF-1 level
Kim et al. (2009)[43]	42	64	67	60	GH ≤ 2.5 μg/L, GH ≤ 1 μg/L (OGTT), normal IGF-1 level
Gondim et al. (2010)[44]	67	75	86	72	GH < 1 μg/L (OGTT), normal IGF-1 level
Jane et al. (2011)[45]	60	70	100	61	GH ≤ 1.0 μg/L ± GH ≤ 1.0 μg/L (OGTT), normal IGF-1 level
Wagenmakers et al. (2011)[46]	36	56	56	N/A	GH < 2 mU/L (OGTT), normal IGF-1 level
Starke et al. (2013)[31]	113	71	85	66	GH ≤ 1.0 μg/L ± GH ≤ 0.4 μg/L (OGTT), normal IGF-1 level

surgery in acromegaly the term "cure" should be used carefully. Even if the most stringent criteria are complied, persisted tumor is revealed by MRI in a few patients[45] and few recurrences occur. Thus, no definite "cure" of the disease can be claimed. Even after "successful" surgical treatment, GH levels remain detectable. Frequency and size of GH-secretion pulses are not normal.[49] The most accurate term to be used in acromegaly is "normalization" of GH secretion and IGF-1 levels, meaning relief of symptoms, restoration of a normal life expectancy, and a low risk of recurrence of the disease. Since IGF-1 levels measured in different assays cannot be compared because of the changing IGF-1 assays over time, the real value of postoperative IGF-1 levels as indicators of long-term outcome are not yet clearly established. Future development of more reliable assays,[9] might hopefully allow a better definition of "cure" in acromegalic patients. Assuredly, even the most sophisticated imaging is not as sensitive as the parameters of GH and IGF-1. Thus, a patient with normal postoperative MR (Fig. 3B) can show normalized GH and IGF-1 levels, respectively or suffer from persistent disease. In contrast, depictions of residual tumor like parasellar tumor portions lateral of the carotid artery are always associated with active acromegaly (Fig. 4B).

In the current literature, mostly the "normalization" criteria established at the Cortina d'Ampezzo Consensus Conference are used.[11] The overall rate for endocrinological remission rates with respect to normalization of

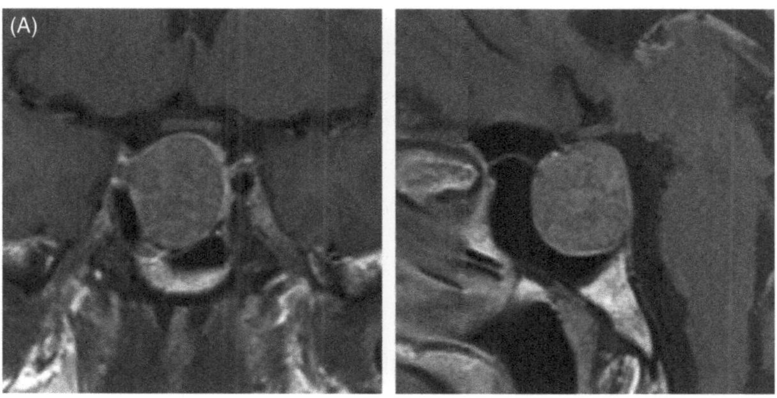

Figure 3 (A) An enclosed intra- and suprasellar pituitary adenoma presents with preoperative GH levels of 14.3 ng/mL in a 55-year-old male patient on the coronal and sagittal T1-weighted MR images. (B) Postoperative imaging demonstrates the ideal result after transsphenoidal resection of a large pituitary adenoma on corresponding sections, namely cisternal herniation (partially "empty sella"). Correspondingly, a normal postoperative IGF-1 level is found 126.2 ng/mL (GH level 1.2 ng/mL).

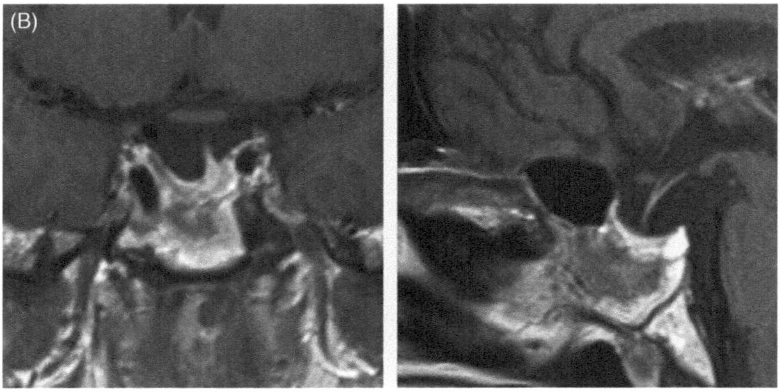

Figure 3 (Continued)

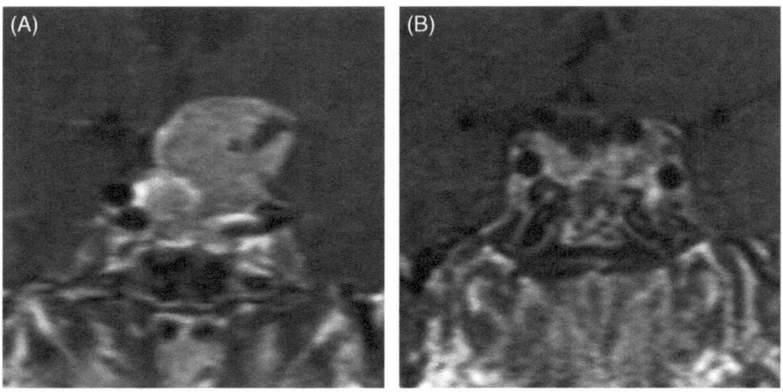

Figure 4 (A) A large intra-, supra-, and parasellar invasive pituitary adenoma presents with high preoperative GH levels (29.5 ng/mL) in a 59-year-old male patient on coronal T1-weighted MR images. (B) Postoperative MR imaging after transsphenoidal and transcranial resection demonstrates an intra-, supra-, and bilateral (predominantly left) parasellar residual tumor which is associated with persistent acromegaly and only slightly lower GH levels (postoperative GH 20.1 ng/mL).

basal GH (below 2.5 μg/L), normalization of IGF-1, and suppression of GH levels below 1 μg/L following oral glucose load was found to range between 42 and 82% (Table 2). Tumor-specific factors, like tumor size, invasive character, tumor extension, and secretory activity were found as predictors of the surgical result. In the current Clinical Practice Guideline established by the Endocrine Society[12] an undetectable GH value together with normal IGF-1 levels after surgery are considered sufficient for indicating remission.

Table 2 Results of Reoperations for GH-Secreting Pituitary Adenomas in Acromegaly Using Different Criteria for Cure of the Disease: A Literature Survey.

Authors	OP/ Cured	Normalization Rate in %	Criteria for Remission
Grisoli et al. (1985)[13]	7/0	0	Basal GH < 5 μg/L
Losa et al. (1989)[38]	3/0	0	Basal GH < 2.5 μg/L + GH < 1 μg/L (OGTT) + normal IGF-1 level
Fahlbusch et al. (1992)[32]	90/16	18	Basal GH < 2.5 μg/L + GH < 2 μg/L (OGTT) ± normal IGF-1 level
Long et al. (1996)[50]	16/3	19	Basal GH < 2.5 μg/L ± GH < 2 μg/L (OGTT) ± normal IGF-1 level
Abe and Lüdecke (1998)[51]	28/16	57	Basal GH < 4.5 μg/L + GH < 2 μg/L (OGTT)
Freda et al. (1998)[41]	10/2	20	Basal GH < 2.5 μg/L + GH < 2 μg/L (OGTT) ± normal IGF-1 level
Shimon et al. (2001)[52]	12/1	8	Basal GH < 2.5 μg/L + GH < 1 μg/L (OGTT) ± normal IGF-1 level
Kurosaki et al. (2003)[53]	22/13	59	Basal GH < 2.0 μg/L + normal IGF-1 level
Yamada et al. (2010)[54]	53/31	59	Basal GH < 1.0 μg/L ± GH < 1 μg/L (OGTT) + normal IGF-1 level
Wilson et al. (2013)[55]	14/8	57	Basal GH < 1.0 μg/L ± GH < 0.4 μg/L (OGTT) + normal IGF-1 level

Postoperative GH levels < 1 μg/L are considered "control" of the disease and correlate with normalization of the mortality risk.

Concerning tumor size, microadenomas have a more favorable outcome than macroadenomas, particularly those with extrasellar extension. In microadenomas, a remission rate as high as 100% has been reported.[45] With increasing tumor diameter and grade of extrasellar extension, remission rates decrease continuously to reach 0% for giant adenomas. Outcome analyses series of Ross *et al.*[15] and Tindall *et al.*[16] found similar results. The main reason for the decrease in success in larger adenomas is the higher incidence of invasive tumors increasing with the size of the adenoma. Pituitary adenomas are considered invasive if they have infiltrated or perforated the sellar diaphragm, the basal dura, the clivus or the sphenoidal, and cavernous sinuses as the normal anatomical limits of the sella.[56] Although invasion tends to be more common with increasing tumor size, also microadenomas may show an invasive character and, in contrast, large tumors may also reach a reasonable size by simply displacing anatomical structures without actually invading them. However, localized or generalized invasion must not be equated with malignancy of the tumor. Pituitary carcinoma is a diagnosis which is only exceptionally justified when subarachnoid dissemination, brain or distant metastases occur. Pituitary carcinomas associated with acromegaly are extremely rare.[57] Invasion can be documented by intraoperative findings, histological examination, and/or neuroradiological investigations. Surgical invasion describes the invasive nature of the tumor as seen intraoperatively. In contrast, the demonstration of histological invasion, although more frequent, is only possible by investigating the anatomical structures in the direct vicinity of the tumor. Dural invasion can be detected in up to 80% of the cases, if specifically investigating the basal dura.[26,57] In our most recent series,[20] the overall remission rate of noninvasive adenomas was 76% and dropped to 24% for invasive tumors. A normalization rate of 5.6% of acromegalic patients with invasive tumors was reported by Mortini *et al.*[33] Less data are available for transcranial resection of pituitary adenomas. Tumors less likely to achieve remission are often included into the transcranial series, resulting in a large selection bias. We thus observed a remission rate of only 5.2% following transcranial surgery.[20]

The response to surgical therapy seems also to be influenced by the secretory activity of the adenoma, as reported in several published series.[17,19,40] Different thresholds of preoperative basal GH levels between 40 and 70 μg/L have been proposed to be associated with a poor prognosis. The higher preoperative measured GH levels exceed these values the lower the likelihood of postoperative biochemical remission. Several years ago, we reported an almost linear character of this inverse relationship[32] and recently confirmed these findings with updated figures.[20] If defining an arbitrary threshold for the secretory activity of an adenoma with a basal GH level of 50 μg/L, the overall cure rate drops from 74% in cases with preoperative GH levels below 50 μg/L to 25% in cases with preoperative GH levels above 50 μg/L.[53]

There is conflicting data from randomized controlled studies discussing whether presurgical treatment with somatostatin analogs would improve surgical results in acromegaly.[58–60] While Kristof *et al.*[58] as well as Plöckinger and Quabbe[60] reported no differences in the outcome, others[61–64] suggested advantage in the pretreated cohort. Noteworthy, an advantage is only seen in the series in which the surgical results without medical pretreatment are relatively poor.[62–64] In one study, there was a short-term, but no long-term benefit.[63]

"Debulking" of the tumor, however, seems to result in some benefit, even in patients in whom hormonal normalization cannot be achieved by surgery. Those operated patients showed a better response to treatment with somatostatin analogs.[21–24] Additionally, a reduction of tumor volume allows lower radiation exposure to critical structures like visual pathways and brainstem in patients who need to undergo irradiation treatments postoperatively.

Another important factor influencing surgical success (independent from the tumor's characteristics) is the surgeon's experience. Within one center, with gaining experience over time the surgical success rate increases and the failure rate decreases.[35] If the number of pituitary surgeons is reduced in a department, the "expert pituitary surgeon" can gain more individual experience following increased surgical success rates.[65,66] Countrywide registry studies reported enormously differing results of surgery.[67] Additionally, the results reached in the entire healthcare system are worse than those being published by specialized experts.[68] However, even in these surveys an impressive number of patients reach the goals of acromegaly treatment after surgery.

6. COMPLICATIONS

To date, complication rates of both transsphenoidal and transcranial operations are relatively low in expert centers.[33] The endocrinological result complication rates are also clearly related to the surgeon's experience and the case load of the services. In analyzing health insurance data Barker *et al.*[69] have investigated the relationship between the total number of pituitary operations performed by individual surgeons and the annual number of cases done within one service to both mortality and complication rate. They were able to show a clearly inverse relationship of the complication rates to both hospital and surgeon volume. Although these data were derived from various types of pituitary adenomas, they are also applicable to growth hormone (GH) secreting tumors. Reviewing the operative complications in 30 published surgical series, Ross and Wilson estimated an operative mortality rate of 1.04%.[15] In our most recent publication on acromegalic patients, one patient died (0.1%) due to hemorrhage in the suprasellar portion of an only partially resected giant adenoma via the transsphenoidal route.[20] Additionally, morbidity was relatively low: meningitis and CSF-fistula occurred in 1.8 and 0.8% of the cases, respectively. Thus, the complication rate is comparable to that of nonacromegalic patients undergoing transsphenoidal surgery for all kinds of pituitary adenomas, due to specific precautions taken in the high-risk group of acromegalic patients. Worsening of pituitary function was only rarely found following primary transsphenoidal surgery.[20,33] Additional endocrine deficiencies requiring permanent replacement therapy were only found in 16 of our cases (3.2%).[20] Preoperative evaluation should include assessment of cardiac function, blood pressure, respiratory function, and metabolic disturbances. Morphological abnormalities (i.e., macroglossia and enlargement of the jaw) require a careful anesthesiological evaluation as they may hinder endotracheal intubation.[28] Fiberoptic endotracheal intubation is necessary in several patients. Furthermore, patients with severe sleep apnea should be monitored in an intensive care unit after surgery. For keeping complication rates as low as possible, a careful management of all these problems is needed throughout the perioperative period. If a patient's general condition is markedly reduced, presurgical short-term treatment with somatostatin analogs should be considered.[59] The patient's condition might be improved and the complication rate might be decreased by that procedure, although there is no evidence for this concept.

7. RECURRENCES

With having clearly defined success or failure following surgical treatment of secreting tumors, persistent disease means failure to achieve the currently used normalization criteria. Only patients who once reach normalized GH secretion and IGF-1 levels may develop recurrent disease. Long-term persistent normalization is the goal of treatment. Normalization of GH secretion as defined by consensus conferences[11] refers to the conditions found in the normal population. Generally, postoperative normalization is paralleled by a normal (delayed postoperative) MRI investigation (Fig. 3)[17] although, particularly in the early postoperative phase, biochemical remission and radiological findings are not always necessarily congruent. Occasionally normal GH and IGF-1 levels can be found in patients with suspected residual tumor.

Unfortunately, the medical literature provides surprisingly inhomogeneous recurrence rates after transsphenoidal surgery. Obviously this depends on the follow-up periods and, of course, on remission criteria. Remarkably, remission criteria were not constantly congruent with today's perception. The reported recurrence rates vary from 0 to 16%[70] even when basal GH levels were just below 5 μg/L within 5–10 years after surgery.

These findings certainly do not reflect remission according to the most recent criteria and, from our present understanding, patients with persistent disease are included in this statistics. A low recurrence incidence has been reported in the follow-up of patients showing suppression of GH levels to below 2 μg/L following oral glucose load.[36,70] Recurrence rates between 0[47,71] and 8.4%[3] have been reported if using the criteria of the consensus conference on normalization of acromegaly after treatment.[11] In a recent reevaluation of our own large series, 0.4% real recurrences could be found in 368 patients after follow-up for a mean of 10 years.[20] These findings show that real recurrences are rare events in acromegaly once currently used criteria of normalization are reached. However, in other series with less-stringent criteria of surgical remission reported recurrence rates reach up to 17.8%.[39]

Recurrent acromegaly may develop early or late. Latencies of 10 years and more have been described.[72] Particularly after having performed a GH-suppressing medication for a long time, a rebound effect can occur after withdrawal of preoperative somatostatin treatment. This may be a major confounder for "early recurrence" in acromegaly. Remarkably, recurrence is not always necessarily associated with corresponding reappearance of tumor in MR-imaging. A rough estimation would suggest reappearance of visible tumor in slightly less than 50% of patients with true recurrent acromegaly.[73]

As prognostic factors suggested by several investigators for the most favorable prognosis for long-term remission and a low risk for recurrent acromegaly are very low postoperative basal GH levels associated with very low GH levels during suppressing and stimulating tests following surgery. Using a highly sensitive GH assay, Freda *et al.*[74] differentiated the patients in remission into groups with low GH-matching levels of normal controls and patients with still sufficient, but slightly higher suppressed levels. While in all patients with very low GH levels long-term remission was reported, 5 of the 19 patients with suppressed GH levels between 0.14 and 1 μg/L developed recurrences. In a further study,[75] different mean hourly GH levels and different responses to arginine stimulation were reported for the two groups. Biermasz *et al.*[76] also found a persistent postoperative GH increase after TRH to be a prognostic indicator for potential later recurrence. Accordingly, these findings would suggest very low postoperative basal, stimulated, and suppressed GH levels after surgical treatment as indicators for a low risk of recurrence. As a consequence, higher although still "normal" levels might indicate small tumor residuals, whose progression can later establish recurrent acromegaly. Ronchi *et al.*[71] also grouped their postoperative patients into one group with very low GH levels (matching normal controls) and another group with higher GH levels, but still below 1 μg/L after oral glucose load. In contrast, both these groups showed no difference after follow-up. Both clinical parameters and signs of disease activity were similar and in none of the groups recurrences occurred.

Apart from hormonal parameters there are other prognostic factors for the long-term outcome, such as postoperative radiotherapy, normal postoperative MRI, and a noninvasive tumor. Postoperative radiotherapy may be a preventive measure against later recurrence. Biermasz *et al.*[72] reported 5 recurrences in 27 patients who did not undergo postoperative irradiation. In contrast, no recurrence was observed in the 9 patients who underwent postoperative radiotherapy.

There is limited experience with the therapy of recurrent acromegaly. Treatment options consist of repeat surgery, medical treatment with somatostatin analogs, radiotherapy, and further observation. There are limited data on repeat surgery in recurrent acromegaly. If considering the results of repeat transsphenoidal surgery for residual and recurrent tumor together, a remission rate of slightly below 20% can be anticipated, irrespective of radiological findings.[32,41,50] Abe and Lüdecke[51,77] and Kurosaki *et al.*[53] report on considerably better results after transsphenoidal reoperations in one center. Since real recurrences are rare events, there are hardly data to document if medical treatment or irradiation in case of recurrence is as efficient as it is in the primary treatment of acromegaly or in persistent disease.

8. PERSISTENT DISEASE

Management of patients with persistent disease after initial surgery is a matter of controversy. For the determination, which patient is a good candidate for repeat surgery, features like the magnitude of the GH and IGF-1 excess and the size and shape of residual tumor as seen on the postoperative MR images can be used. Reexploration might be particularly successful in those cases in which neither surgical nor histological invasion was demonstrated during the initial surgery which was not performed by a particularly experienced surgeon. Formation of scar tissue and loss of anatomical landmarks for midline orientation make repeat surgeries technically more demanding. In our own series, reexplorations have been performed in 156 cases, mainly in patients with MRI studies that demonstrated residual adenoma. Expectedly, overall remission rate after repeat surgery was much worse (21.3%) than in the initial surgery group. Nevertheless, in patients with noninvasive tumors and an initial GH level below 40 μg/L prior to the first surgical approach, a remission rate of 37.3% could be achieved.[20] For evaluation of the effectiveness of a secondary transsphenoidal surgery, 28 patients with persistent or recurrent acromegaly were analyzed by Abe *et al.*[51] An overall endocrinological remission rate of 57.1% (16 of 28 patients) without serious morbidity or mortality has been reported. Absence of major surgical complications and worsening of other pituitary functions combined with the possibility of long-term control of the disease support reexploration as a serious treatment option for selected patients with persistent acromegaly.

9. NEW TECHNOLOGIES

Novel surgical techniques or variants of established procedures have been enabled by new technical possibilities.[25]

The approach to parasellar lesions is made safer by use of an endoscope with its main advantage of panoramic visualization of the anatomical structures through the wide angle of view and powerful light devices. Furthermore, the direct nasal approach without paraseptal dissection as it is used for endoscopic pituitary surgery, may often avoid the need for postoperative insertion of nasal tamponades and may therefore be beneficial to a better subjective feeling for the patient due to free nasal ventilation.[29,30]

Another attempt to improve the surgical results by identifying residual tumor was the performance of intraoperative blood sampling and determination of GH levels.[77] Ideally, a rapid extinction of GH to very low levels can be documented within several hours after successful transsphenoidal surgery. An intraoperative decrease is suggested to be used to predict the extent of resection. A major disadvantage is that this technique is time consuming and requires laboratory facilities for rapid intraoperative GH assessments.

Neuronavigation is more and more used for the intraoperative localization of anatomic structures. Major advantages are a pointer-based localization and the possibility of virtual surgery with segmentation and three-dimensional reconstruction of the tumor and its adjacent structures from the preoperative MRI data. Superimposing the tumor margins and the carotid arteries allows for a safe, straightforward approach to the sella even in cases with vascular anomalies, narrow sellar floors, poor pneumatization of the paranasal sinuses, and loss of anatomical landmarks in reoperations.[6,7,25]

Intraoperative MR imaging is also increasingly available in all countries. It allows neurosurgeons to perform surgery interactively using MRI guidance and, more importantly, allows intraoperative estimation of the amount of tumor resection, which is particularly valuable in large intra-, supra-, and parasellar tumors. From morphological considerations intraoperative MR is the ideal quality control to immediately determine the extent of adenoma resection and identify residual tumor tissue.[62,78–80] Its application during transsphenoidal surgery of acromegalic patients with complex adenomas has been strongly recommended[78,80–83] since the rate of total tumor resections was shown to be increased.

REFERENCES

1. Orme SM, McNally RJ, Cartwright RA, Belchetz PE. Mortality and cancer incidence in acromegaly: a retrospective cohort study. United Kingdom Acromegaly Study Group. *J Clin Endocrinol Metab.* 1998;83:2730–2734.
2. Bates AS, Van't Hoff W, Jones JM, Clayton RN. An audit of outcome of treatment in acromegaly. *Quart J Med.* 1993;86:293–299.
3. Beauregard C, Truong U, Hardy J, Serri O. Long-term outcome and mortality after transsphenoidal adenomectomy for acromegaly. *Clin Endocrinol (Oxf).* 2003;58:86–91.
4. Rajasoorya C, Holdaway IM, Wrightson P, Scott DJ, Ibbertson HK. Determinants of clinical outcome and survival in acromegaly. *Clin Endocrinol (Oxf).* 1994;41:95–102.
5. Swearingen B, Barker FG, Katznelson L, Biller BM, Greenspoon S, Klibanski A, Moayeri N, Black PM, Zervas NT. Long-term mortality after transsphenoidal surgery and adjunctive therapy for acromegaly. *J Clin Endocrinol Metab.* 1998;83:3419–3426.
6. Buchfelder M, Fahlbusch R. The "classic" transsphenoidal approach for resection of pituitary tumors. *Operat Techn Neurosurg.* 2002;4:210–217.
7. Couldwell WT. Transsphenoidal and transcranial surgery for pituitary adenomas. *J Neurooncol.* 2004;69:237–256.
8. Coloa A, Pivonello R, Auriemma RS, Galdiero M, Guerra E, Milone F, De Leo M, Lombardi G. New perspectives in the medical treatment of acromegaly. *J Endocrinol Invest.* 2005;28:58–66.
9. Melmed S, Casanueva F, Cavagnini F, Chanson P, Frohman LA, Gaillard R, Ghigo E, Ho K, Jaquet P, Kleinberg D, Lamberts S, Laws E, Lombardi G, Sheppard MC, Thorner M, Vance ML, Wass JA, Giustina A. Consensus statement: medical management of acromegaly. *Eur J Endocrinol.* 2005;153:737–740.
10. Melmed S, Casanueva FF, Cavagnini F, Chanson P, Frohman L, Grosman A, Ho K, Kleinberg D, Lamberts S, Laws E, Lombardi G, Vance ML, von Werder K, Wass J, Giustina A. Guidelines for acromegaly management. *J Clin Endocrinol Metab.* 2002;87: 4054–4058.

11. Giustina A, Barkan A, Casanueva FF, Cavagnini F, Frohman L, Ho K, Veldhuis J, Wass J, Von Werder K, Melmed S. Criteria for cure of acromegaly: a consensus statement. *J Clin Endocrinol Metab.* 2000;85:526–529.
12. Katznelson L, Laws ER, Melmed S, Molitch ME, Murad MH, Wass JAH. Acromegaly: an endocrine society clinical practice guideline. *J Clin Endocrinol Metab.* 2014;99:3933–3951.
13. Grisoli F, Leclercq T, Jaquet P, Guibout M, Winteler JP, Hassoun J, Vincentelli F. Transsphenoidal surgery for acromegaly—long term results in 100 patients. *Surg Neurol.* 1985;23:513–519.
14. Hardy J. Transsphenoidal microsurgery of the pathological and normal pituitary. *Clin Neurosurg.* 1969;16:185–217.
15. Ross DA, Wilson CB. Results of transsphenoidal microsurgery for growth hormone-secreting pituitary adenoma in a series of 214 patients. *J Neurosurg.* 1988;68:854–867.
16. Tindall GT, Oyesiku NM, Watts NB, Clark RV, Christy JH, Adams DA. Transsphenoidal adenomectomy for growth hormone-secreting pituitary adenomas in acromegaly: outcome analysis and determinants of failure. *J Neurosurg.* 1993;78:205–215.
17. Zervas NT. Multicenter surgical results in acromegaly. In: Ludecke D, Tolis G, eds. In: Growth Hormone, Growth Factors and Acromegaly. New York: Raven Press; 1987:
18. Kreutzer J, Vance ML, Lopes MB, Laws Jr ER. Surgical management of GH-secreting pituitary adenomas: an outcome study using modern remission criteria. *J Clin Endocrinol Metab.* 2001;86:4072–4077.
19. Laws ER, Vance ML, Thapar K. Pituitary surgery for the management of acromegaly. *Horm Res.* 2000;53:71–75.
20. Nomikos P, Buchfelder M, Fahlbusch R. The outcome of surgery in 668 patients with acromegaly using current criteria of biochemical 'cure'. *Eur J Endocrinol.* 2005;152: 379–387.
21. Colao A, Attanasio R, Pivonello R, Cappabianca P, Cavallo L, Lasio G, Lodrini A, Lombardi G, Gozzi R. Partial surgical removal of growth hormone-secreting pituitary tumors enhances the response to somatostatin analogs in acromegaly. *J Clin Endocrinol Metab.* 2006;91:85–92.
22. Jallad RS, Musolino NR, Kodaira S, Cescato VA, Bronstein MD. Does partial surgical removal influence the response to octreotide-LAR in acromegalic patients previously resistant to the somatostatin analogue? *Clin Endocrinol.* 2007;67:310–315.
23. Karavitaki N, Turner HE, Adams CBT, Cudlip S, Byrne JV, Fazal-Sanderson V, Rowlers S, Trainer PJ, Wass JAH. Surgical debulking of pituitary macroadenomas causing acromegaly improves control by lanreotide. *Clin Endocrinol.* 2008;68:970–975.
24. Petrossians P, Borges-Martins L, Espinoza C, Daly A, Betea D, Valdes-Sorcin H, Stevenaert A, Chanson P, Beckers A. Gross total resection or debulking of pituitary adenomas improves hormonal control of acromegaly by somatostatin analogs. *Eur J Endocrinol.* 2005;152:61–66.
25. Buchfelder M. Treatment of pituitary tumors: surgery. *Endocrine.* 2005;28:67–75.
26. Ahmadi J, North CM, Segall HD, Zee CS, Weiss MH. Cavernous sinus invasion by pituitary adenomas. *AJR Am J Roentgenol.* 1986;146:257–262.
27. Bonadonna S, Doga M, Gola M, Mazziotti G, Giustina A. Diagnosis and treatment of acromegaly and its complications: consensus guidelines. *J Endocrinol Invest.* 2005;28: 43–47.
28. Schmitt H, Buchfelder M, Radespiel-Troger M, Fahlbusch R. Difficult intubation in acromegalic patients: incidence and predictability. *Anesthesiology.* 2000;93:110–114.
29. Cappabianca P, Cavallo LM, de Divitis E. Endoscopic endonasal transsphenoidal surgery. *Neurosurgery.* 2004;55:933–940.
30. Rudnick A, Kos-Kurdla B, Larysz D, Zawadzki T, Bazowski P. Endoscopic transsphenoidal treatment of hormonally active pituitary adenomas. *Neuro Endocrinol Lett.* 2007;28:438–444.

31. Starke RM, Raper DMS, Payne SC, Vance ML, Oldfield EH, Jane JA. Endoscopic vs microsurgical transsphenoidal surgery for acromegaly: outcomes in a concurrent series of patients using modern criteria for remission. *J Clin Endocrinol Metab.* 2013;98: 3190–3198.
32. Fahlbusch R, Honegger J, Buchfelder M. Surgical management of acromegaly. *Endocrinol Metab Clin North Am.* 1992;21:669–692.
33. Mortini P, Losa M, Barzaghi R, Boari N, Giovanelli M. Results of transsphenoidal surgery in a large series of patients with pituitary adenomas. *Neurosurgery.* 2005;56: 1222–1233.
34. Abosch A, Tyrrell JB, Lamborn KR, Hannegan LT, Applebury CB, Wilson CB. Transsphenoidal microsurgery for growth hormone-secreting pituitary adenomas: initial outcome and long-term results. *J Clin Endocrinol Metab.* 1998;83:3411–3418.
35. Ahmed S, Elsheikh M, Stratton IM, Page RC, Adams CB, Wass JA. Outcome of transphenoidal surgery for acromegaly and its relationship to surgical experience. *Clin Endocrinol (Oxf).* 1999;50:561–567.
36. Buchfelder M, Brockmeier S, Fahlbusch R, Honegger J, Pichl J, Manzl M. Recurrence following transsphenoidal surgery for acromegaly. *Horm Res.* 1991;35:113–118.
37. De P, Rees DA, Davies N, John R, Neal J, Mills RG, Vafidis J, Davies JS, Scanlon MF. Transsphenoidal surgery for acromegaly in Wales: results based on stringent criteria of remission. *J Clin Endocrinol Metab.* 2003;88:3567–3572.
38. Losa M, Oeckler R, Schopohl J, Muller OA, Alba-Lopez J, von Werder K. Evaluation of selective transsphenoidal adenomectomy by endocrinological testing and somatomedin-C measurement in acromegaly. *J Neurosurg.* 1989;70:561–567.
39. Davis DH, Laws Jr ER, Ilstrup DM, Speed JK, Caruso M, Shaw EG, Abboud CF, Scheithauer BW, Root LM, Schleck C. Results of surgical treatment for growth hormone-secreting pituitary adenomas. *J Neurosurg.* 1993;79:70–75.
40. Sheaves R, Jenkins P, Blackburn P, Huneidi AH, Afshar F, Medbak S, Grossman AB, Besser GM, Wass JA. Outcome of transsphenoidal surgery for acromegaly using strict criteria for surgical cure. *Clin Endocrinol (Oxf).* 1996;45:407–413.
41. Freda PU, Wardlaw SL, Post KD. Long-term endocrinological follow-up evaluation in 115 patients who underwent transsphenoidal surgery for acromegaly. *J Neurosurg.* 1998;89:353–358.
42. Frank G, Pasquini E. Endoscopic endonasal cavernous sinus surgery, with special reference to pituitary tumors. *Front Horm Res.* 2006;34:64–82.
43. Kim MS, Jang HD, Kim OL. Surgical results of growth hormone-secreting pituitary adenoma. *J Korean Neurosurg Soc.* 2009;45:271–274.
44. Gondim JA, de Almeida JP, Albuquerque LAF, Games E, Schops M, Ferraz T. Pure endoscopic transsphenoidal surgery for treatment of acromegaly: results of 67 cases treated in a pituitary center. *Neurosurg Focus.* 2010;29:E7.
45. Jane JA, Starke RM, Elzoghby MA, Reames DL, Payne SC, Thorner MO, Marshall JC, Laws ER, Vance ML. Endoscopic transsphenoidal surgery for acromegaly: remission using modern criteria, complications and predictors of outcome. *J Clin Endocrinol Metab.* 2011;96:2732–2740.
46. Wagenmakers MAEM, Netea-Maier RT, van Lindert EJ, Pieters GFFM, Grotenhuis AJA, Hermus ARMM. Results of endoscopic transsphenoidal pituitary surgery in 40 patients with a growth hormon-secreting macroadenoma. *Acta Neurochir.* 2011;153: 1391–1399.
47. Valdemarsson S, Bramnert M, Cronquist S, Elner A, Eneroth CM, Hedner P, Lindvall-Axelsson M, Nordstrom CH, Stromblad LG. Early postoperative basal serum GH level and the GH response to TRH in relation to the long-term outcome of surgical treatment for acromegaly: a report on 39 patients. *J Intern Med.* 1991;230:49–54.

48. Valdemarsson S, Ljunggren S, Bramnert M, Norrhamn O, Nordstrom CH. Early post-operative growth hormone levels: high predictive value for long-term outcome after surgery for acromegaly. *J Intern Med.* 2000;247:640–650.
49. Peacey SR, Toogood AA, Veldhuis JD, Thorner MO, Shalet SM. The relationship between 24-hour growth hormone secretion and insulin-like growth factor I in patients with successfully treated acromegaly: impact of surgery or radiotherapy. *J Clin Endocrinol Metab.* 2001;86:259–266.
50. Long H, Beauregard H, Somma M, Comtois R, Serri O, Hardy J. Surgical outcome after repeated transsphenoidal surgery in acromegaly. *J Neurosurg.* 1996;85:239–247.
51. Abe T, Lüdecke DK. Recent results of secondary transnasal surgery for residual or recurring acromegaly. *Neurosurgery.* 1998;42:1013–1021.
52. Shimon I, Cohen ZR, Ram Z, Handani M. Transsphenoidal surgery for acromegaly: endocrinological follow-up of 98 patients. *Neurosurgery.* 2001;48:1239–1243.
53. Kurosaki M, Luedecke DK, Abe T. Effectiveness of secondary transnasal surgery in GH-secreting pituitary macroadenomas. *Endocr J.* 2003;50:635–642.
54. Yamada S, et al. Repeat transsphenoidal surgery for the treatment of remaining or recurring pituitary tumors in acromegaly. *Neurosurgery.* 2010;67:949–956.
55. Wilson TJ, McKean EL, Barkan AL, Chandler WF, Sullivan SE. Repeat endoscopic transsphenoidal surgery for acromegaly: remission and complications. *Pituitary.* 2013;16: 459–464.
56. Buchfelder M, Fahlbusch R, Adams EF, Kiesewetter F, Thierauf P. Proliferation parameters for pituitary adenomas. *Acta Neurochir Suppl.* 1996;65:18–21.
57. Kaltsas GA, Nomikos P, Kontogeorgos G, Buchfelder M, Grossman AB. Clinical review: diagnosis and management of pituitary carcinomas. *J Clin Endocrinol Metab.* 2005;90: 3089–3099.
58. Kristof RA, Stoffel-Wagner B, Klingmuller D, Schramm J. Does octreotide treatment improve the surgical results of macro-adenomas in acromegaly? A randomized study. *Acta Neurochir.* 1999;141:399–405.
59. Colao A, Ferone D, Cappabianca P, del Basso De Caro ML, Marzullo P, Monticelli A, Alfieri A, Merola B, Cali A, de Divitiis E, Lombardi G. Effect of octreotide pretreatment on surgical outcome in acromegaly. *J Clin Endocrinol Metab.* 1997;82: 3308–3314.
60. Plöckinger U, Quabbe HJ. Presurgical octreotide treatment in acromegaly: no improvement of final growth hormone (GH) concentration and pituitary function. A long-term case-control study. *Acta Neurochir (Wien).* 2005;147:485–493.
61. Carlsen SM, Lund-Johansen M, Schreiner T, Aanderud S, Johannesen Øsah, Svartberg J, Cooper JG, Hald JK, Fougner SL, Bollerslev J. Preoperative octreotide treatment in newly diagnosed acromegalic patients with macroadenomas increases cure short-term postoperative rates: a prospective, randomized trial. *J Clin Endocrinol Metab.* 2008;93: 2984–2990.
62. Fahlbusch R, Thapar K. New developments in pituitary surgical techniques. *Baillieres Best Pract Res Clin Endocrinol Metab.* 1999;13:471–484.
63. Mao ZG, Zhu YH, Tang HL, Wang DY, Zhou J, He DS, Lan H, Luo BN, Wang HJ. Preoperative lanreotide treatment in acromegalic patients with macroadenomas increases short-term postoperative cure rates: a prospective, randomized trial. *Eur J Endocrinol.* 2010;162:661–666.
64. Shen M, Shou X, Wang Y, Zhang Z, Wu J, Mao Y, Li S, Zhao Y. Effect of presurgical long-acting octreotide treatment in acromegaly patients with invasive pituitary macroadenomas: a prospective randomized study. *Endocr J.* 2010;57:1035–1044.
65. Clayton RN, Stewart PM, Shalet SM, Wass JA. Pituitary surgery for acromegaly should be done by specialists. *BMJ.* 1999;319:588–589.

66. Lissett CA, Peacey SR, Laing I, Tetlow L, Davis JRE, Shalet S. The outcome of surgery for acromegaly: the need for specialist pituitary surgeon for all types of growth hormone (GH) secretion. *Clin Endocrinol.* 1998;49:653–657.
67. Bates PR, Carson MN, Trainer PJ, Wass JAH. The UK National Acromegaly Register Study Group.Wide variation in surgical outcomes for acromegaly in the UK. *Clin Endocrinol.* 2008;68:136–142.
68. Petersenn S, Buchfelder M, Reincke M, Strasburger CJ, Franz H, Lohmann R, Quabbe HJ, Plöckinger U. Participants of the German Acromegaly Register. Results of surgical and somatostatin analog therapies and their combination in acromegaly: a retrospective analysis of the German Acromegaly Register. *Eur J Endocrinol.* 2008;159:525–532.
69. Barker FG, Klibanski A, Swearingen B. Transsphenoidal surgery for pituitary tumors in the United States, 1996–2000: mortality, morbidity, and the effects of hospital and surgeon volume. *J Clin Endocrinol Metab.* 2003;88:4709–4719.
70. Arafah BM, Rosenzweig JL, Fenstermaker R, Salazar R, McBride CE, Selman W. Value of growth hormone dynamics and somatomedin C (insulin-like growth factor I) levels in predicting the long-term benefit after transsphenoidal surgery for acromegaly. *J Lab Clin Med.* 1987;109:346–354.
71. Ronchi CL, Varca V, Giavoli C, Epaminonda P, Beck-Peccoz P, Spada A, Arosio M. Long-term evaluation of postoperative acromegalic patients in remission with previous and newly proposed criteria. *J Clin Endocrinol Metab.* 2005;90:1377–1382.
72. Biermasz NR, van Dulken H, Roelfsema F. Ten-year follow-up results of transsphenoidal microsurgery in acromegaly. *J Clin Endocrinol Metab.* 2000;85:4596–4602.
73. Zirkzee EJ, Corssmit EP, Biermasz NR, Brouwer PA, Wiggers-De Bruine FT, Kroft LJ, Van Buchem MA, Roelfsema F, Pereira AM, Smit JW, Romijn JA. Pituitary magnetic resonance imaging is not required in the postoperative follow-up of acromegalic patients with long-term biochemical cure after transsphenoidal surgery. *J Clin Endocrinol Metab.* 2004;89:4320–4324.
74. Freda PU, Post KD, Powell JS, Wardlaw SL. Evaluation of disease status with sensitive measures of growth hormone secretion in 60 postoperative patients with acromegaly. *J Clin Endocrinol Metab.* 1998;83:3808–3816.
75. Freda PU, Nuruzzaman AT, Reyes CM, Sundeen RE, Post KD. Significance of "abnormal" nadir growth hormone levels after oral glucose in postoperative patients with acromegaly in remission with normal insulin-like growth factor-I levels. *J Clin Endocrinol Metab.* 2004;89:495–500.
76. Biermasz NR, Smit JW, van Dulken H, Roelfsema F. Postoperative persistent thyrotrophin releasing hormone-induced growth hormone release predicts recurrence in patients with acromegaly. *Clin Endocrinol (Oxf).* 2002;56:313–319.
77. Abe T, Lüdecke DK. Recent primary transnasal surgical outcomes associated with intraoperative growth hormone measurement in acromegaly. *Clin Endocrinol (Oxf).* 1999;50:27–35.
78. Bellut D, Hlavica M, Schmid C, Bernays RL. Intraoperative magnetic resonance imaging-assisted transsphenoidal pituitary surgery in patients with acromegaly. *Neurosurg Focus.* 2010;29:E9.
79. Fahlbusch R, Ganslandt O, Buchfelder M, Schott W, Nimsky C. Intraoperative magnetic resonance imaging during transsphenoidal surgery. *J Neurosurg.* 2001;95:381–390.
80. Fahlbusch R, von Keller B, Ganslandt O, Kreutzer J, Nimsky C. Transsphenoidal surgery in acromegaly investigated by intraoperative high-field magnetic resonance imaging. *Eur J Endocrinol.* 2005;153:239–248.
81. Jahangiri A, Wagner J, Han SW, Zygourakis CC, Han SJ, Tran MT, Miller LM, Tom MW, Kunwar S, Belevins LS, Aghi MK. Morbidity of repeat transsphenoidal surgery assessed in more than 1000 operations. *J Neurosurg.* 2014;121:67–74.

82. Marko NF, LaSota E, Hamrahian AH, Weil RJ. Comparative effectiveness review of treatment options for pituitary microadenomas in acromegaly. *J Neurosurg.* 2012;117: 522–538.
83. Morrison KM, Wu Z, Bidlingmaier M, Strasburger CJ. Findings and theoretical considerations on the usefulness of the acid-labile subunit in the monitoring of acromegaly. *Growth Horm IGF Res.* 2001;11(suppl A):S61–S63.

SECTION III

GH Deficiency

CHAPTER SEVEN

Burden of Growth Hormone Deficiency and Excess in Children

Hugo L. Fideleff[1], Hugo R. Boquete, Martha G. Suárez, Miriam Azaretzky
Department of Medicine, Endocrinology Unit, Hospital T. Alvarez, Buenos Aires, Argentina
[1] Corresponding author: e-mail address: hugofideleff@arnet.com.ar.

Contents

Abstract

Longitudinal growth results from multifactorial and complex processes that take place in the context of different genetic traits and environmental influences. Thus, in view of the difficulties in comprehension of the physiological mechanisms involved in the achievement of normal height, our ability to make a definitive diagnosis of GH impairment still remains limited.

There is a myriad of controversial aspects in relation to GH deficiency, mainly related to diagnostic controversies and advances in molecular biology. This might explain the diversity in therapeutic responses and may also serve as a rationale for new "nonclassical" treatment indications for GH. It is necessary to acquire more effective tools to reach an adequate evaluation, particularly while considering the long-term implications of a correct diagnosis, the cost, and safety of treatments.

On the other hand, overgrowth constitutes a heterogeneous group of different pathophysiological situations including excessive somatic and visceral growth.

Progress in Molecular Biology and Translational Science, Volume 138
ISSN 1877-1173
http://dx.doi.org/10.1016/bs.pmbts.2015.10.009

There are overlaps in clinical and molecular features among overgrowth syndromes, which constitute the real burden for an accurate diagnosis.

In conclusion, both GH deficiency and overgrowth are a great dilemma, still not completely solved. In this chapter, we review the most burdensome aspects related to short stature, GH deficiency, and excess in children, avoiding any details about well-known issues that have been extensively discussed in the literature.

ABBREVIATIONS

ALS	Acid labile subunit
BMI	Body mass index
BMP	Bone morphogenetic proteins
CGH	Comparative genomic hybridization
FGF	Fibroblast growth factor
GH	Growth hormone
GHD	Growth hormone deficiency
GHR	Growth hormone receptor
GH-RP	Growth hormone related protein
GnRH	Gonadotropin-releasing hormone
HESX1	Homeobox expressed in ES cells 1
HR-CGH	High-resolution comparative genomic hybridization
HypoCCS	Hypopituitary control and complications study
ID-MS	Isotopic mass spectrometry
IGF	Insulin-like growth factor
IGF1	Insulin-like growth factor 1
IGF2	Insulin-like growth factor 2
IGFALS	Insulin-like growth factor binding protein, acid labile subunit
IGFBP1	Insulin-like growth factor binding protein 1
IGFBP3	Insulin-like growth factor binding protein 3
IGFBP4	Insulin-like growth factor binding protein 4
IGFBPs	Insulin-like growth factor binding proteins
IGFR	Insulin-like growth factor receptor
IRMA	Immunoradiometric sandwich assay
ISS	Idiopathic short stature
iSYS	Automated IDS-iSYS multi-discipline immunoassay system
KIGS/KIMS	A database sponsored by Pfizer
LHX3	LIM homeobox 3
LHX4	LIM homeobox 4
MC4R	Melanocortin-4 receptor
MEN-1	Multiple endocrine neoplasia type 1
MPHD	Multiple pituitary hormone deficiencies
MRI	Magnetic resonance imaging
NSD1	Nuclear receptor binding SET domain protein 1
POU1F1	Pituitary-specific positive transcription factor 1
PROP1	Homeobox protein prophet of PIT-1
Ptx1	Pituitary homeobox 1

Ptx2	Pituitary homeobox 2
RP	Related protein
SAGhE	Santé adult GH Enfant Study
SD	Standard deviation
SDS	Standard deviation score
SGA	Small for gestational age
SHH	Sonic hedgehog homolog
SHOX	Short stature homeobox
SI	International System of Units
SMR	Standardized mortality ratio
SOX2	Sex determining region Y-box 2
SOX3	Transcription factor SOX-3
STAT5B	Signal transducer and activator of transcription 5B
WHO	World Health Organization
Wnt	Wingless-type (gene family of structurally related genes that encode signaling proteins)

1. INTRODUCTION

Height is a highly hereditable trait with estimates of heritability approximately 90%.[1] Recent *genome-wide association studies* of height have discovered over 180 common variants associated with stature.[2] These variants have small effect sizes and collectively explain approximately 10% of the heritability.[2] Although these common variants are robustly associated with height, when studied as a quantitative trait in the general population, it is not known whether these variants have similar associations with stature in individuals at the extreme trails of the height distribution.[3] Although there are multiple possible scenarios, one possible explanation is the existence of rare or low frequency variants with larger effect sizes, which have been proposed to explain a portion of heritability and which may provide novel biological insights to explain several mechanisms that affect height.[4]

Thus, in view of the difficulties in the comprehension of the physiological mechanisms involved in the achievement of normal height, our ability to make a definitive diagnosis of GH impairment still remains limited, and undoubtedly, we need more effective means of reaching an adequate diagnosis, particularly when considering the long-term implications of a correct diagnosis and the cost and safety of a possible treatment.

In this chapter, we will try to review the most burdensome aspects in relation to GH deficiency and excess in children, avoiding any details about well-known issues that have been extensively discussed in the literature.

2. GROWTH HORMONE DEFICIENCY

Short stature is one of the most common causes of referral to pediatric endocrinologists. There is a wide spectrum of growth disorders due to the different etiologies involved: heterogeneity of the population evaluated in some countries (mainly in those with a large and varied immigration), social inequalities, undernutrition, etc.

In addition, different stages are recognized in human beings: prenatal, neonatal, infancy, childhood, adolescence, adulthood, and senescence. However, over the recent years, a transition period has been recognized and given relevance. Transition refers to a broad set of physical and psychological changes arbitrarily defined as starting in late puberty and ending with full maturation.[5,6]

The definition of GHD in children is difficult and remains arbitrary, in view of the fact that there is a continuity between absent and normal GH secretion during childhood. The estimated frequency of growth retardation caused by disturbed secretion of GH ranges from 1:4000 to 1:10000.[7] Thus, GHD from a practical point of view could be considered as complete or severe or as incomplete or partial, according to defined cutoff points. GHD in childhood is characterized by a combination of auxological, clinical, genetic, radiological, metabolic, and hormonal abnormalities, which may be present to a greater or a lesser extent.

There is a myriad of controversial aspects in relation to GHD, mainly related to diagnostic controversies and advances in molecular biology. This might explain the diversity in therapeutic responses and may also serve as rationale for new "nonclassical" treatment indications for GH.

2.1 Clinical Aspects

Elementary as it may seem, the definition of short stature may be controversial, as there is disagreement over the height and growth velocity charts that should be used. Not all countries worldwide have national reference standards available, and the WHO charts are based only on population samples from some selected countries in different continents.[8] Furthermore, other aspects should be currently taken into account, such as immigration in different regions, which may modify growth parameters, since it is only after two generations that immigrants will become assimilated to individuals from their adopted country.[9]

Another critical issue is the fact that a child with GHD may not necessarily develop short stature. This might be explained in the case of patients, whose parents have a tall target height and despite deficient GH secretion, their height is not below 2 SDS. This can also be observed in children with normal or tall stature who, as a result of an organic process, develop severe deficiency and only after a variable time their height decreases below 2 SDS.

It is worth mentioning that short stature includes panhypopituitarism, isolated GHD, Turner syndrome, Noonan syndrome, chronic renal failure, Prader–Willi syndrome, idiopathic short stature (ISS), homeobox-containing gene (SHOX) alterations, and small for gestational age (SGA) among others. The etiology of these childhood disorders may involve normal variations, genetic defects, malnutrition, chronic diseases, endocrine disturbances, and psychosocial disorders (Table 1). Although GHD is not the most frequent cause of short stature, its early recognition prompts the initiation of GH treatment. Thus, these children with growth failure and short stature, may

Table 1 Short Stature: Etiologies.

1. Primary growth disorders
   a. Small for gestational age
   b. Genetic syndromes with typical clinical features (e.g., Turner syndrome, Prader–Willi syndrome, Silver–Russell syndrome, Down syndrome, Noonan syndrome)
   c. Congenital bone dysplasias
2. Secondary growth disorders
   a. Endocrine causes
      – Growth hormone deficiency
      – Cushing syndrome (exogenous or endogenous)
      – Hypothyroidism
      – Inadequate treatment of precocious puberty
   b. Other disorders of the GH–IGF–I axis
      – IGF-I deficiency
      – ALS deficiency
      – IGF-I resistance
   c. Poorly controlled diabetes mellitus
   d. Malabsorption syndrome
   e. Malnutrition
   f. Chronic renal failure
   g. Metabolic disorders
   h. Other systemic disorders (cardiac, pulmonary, liver, intestinal, chronic anaemia, arthritis, etc.)
   i. Psychosocial conditions (emotional deprivation)
   j. Psycological alterations (anorexia nervosa)
   k. Adverse effects of childhood malignancy therapy

achieve optimal height, also allowing the prevention of metabolic and cardiovascular disorders, and improving quality of life.

The term ISS is used to describe children with non-GHD short stature, in which the etiology of the growth abnormality cannot be identified.[10] Anyway, with the advance of molecular genetics, many patients diagnosed with ISS have different etiologies for their short stature including alterations of the GH receptor, and in the postreceptor GH signaling. In these patients with an unclear etiology of short stature, who have normal levels of GH, it is advisable to evaluate the possible underlying molecular abnormalities in order to predict the response to treatment. For that reason, interpretation of the GH/IGF1 axis in ISS can be challenging (Fig. 1). Obviously, this subgroup of patients should be included into a broad spectrum of the genetic causes of short stature, which range from Turner syndrome to dyschondroplasia, and many well known genetic disorders.[4,11] Therefore, from an auxological point of view, children presenting short stature and/or poor growth velocity should be evaluated for different potential pathologies that impact GH and IGF1.

It is worth mentioning that, some years ago, the Growth Hormone Research Society suggested the following criteria in its consensus guidelines to perform the evaluation of GH/IGF1 axis:[12]

- Severe short stature, defined as a height more than 3 SD below the mean
- Height more than 1.5 SD below the midparental height
- Height more than 2 SD below the mean and a height velocity over 1 year more than 1 SD below the mean for chronological age, or a decrease in height SD of more than 0.5 over 1 year in children over 2 years of age
- In the absence of short stature, a height velocity more than 2 SD below the mean over 1 year or more than 1.5 SD sustained over 2 year (this may occur in GHD, presenting in infancy, or in organic acquired GHD)
- Signs indicative of an intracranial lesion
- Signs of multiple pituitary hormone deficiency
- Neonatal signs and symptoms of GH deficiency (mainly hypoglycemia, prolonged jaundice, and microphallus)

During the transition period, clinical reevaluation of patients, who have been diagnosed with deficiency during childhood is mandatory. It should also be noted that a subset of patients may develop GH deficiency during the transition period.[13] This is not a minor issue, as it gives rise to a series of questions about who should be retested, when and how, and eventually, if treatment continuation during adulthood is warranted, given the impact of

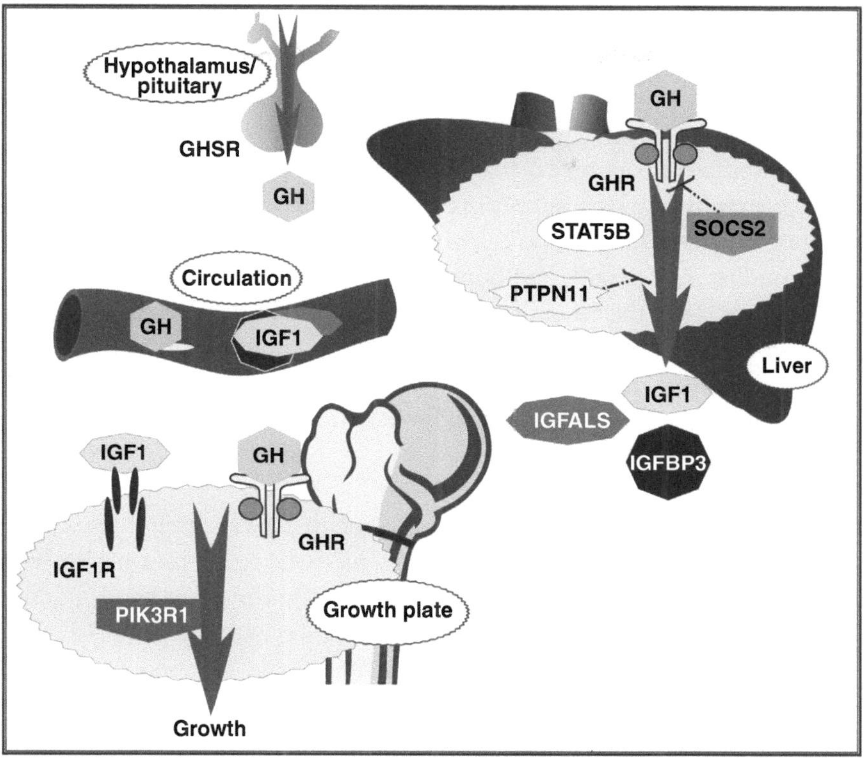

Figure 1 GH/IGF1 axis: schematic representation. Pituitary GH secretion, intracellular hepatic mechanisms for IGF1 production, transportation as a ternary complex in circulation and direct action of IGF1 and GH in growth plate, are depicted. Abbreviations: GH, growth hormone; GHSR, growth hormone secretagogue receptor; GHR, growth hormone receptor; STAT5B, signal transducer and activator of transcription 5B; SOCS2, suppressor of cytokine signaling 2; PTPN11, tyrosine-protein phosphatase nonreceptor type 11; IGF1, insulin-like growth factor 1; IGFALS, IGF acid labile subunit; IGFBP3, insulin-like growth factor binding protein 3; IGF1R, insulin-like growth factor 1 receptor; PIK3R1, phosphatidylinositol 3-kinase regulatory subunit alpha.

GH deficiency on body composition, bone metabolism, cardiovascular risk factors, metabolic profile, and quality of life.

While analyzing the diagnosis of GHD, several confounding factors should be taken into account namely, nutritional status (underweight or overweight children), concomitant medication (e.g., glucocorticoids, psychotropic drugs, antiepileptics), and psychosocial conditions. Regarding obesity in general, obese children are usually taller than their nonobese peers and present higher growth velocity during the prepubertal period.[14] However, some studies have suggested that obese children have subnormal peak response to GH stimulating tests with normal IGF1 levels.[15] This has led to the hypothesis that some obese children could be overdiagnosed, contributing to an increase in the higher prevalence of false positive diagnoses of GHD. Nevertheless, the effect of adiposity on growth velocity and other auxological parameters has not been fully studied. Moreover some authors demonstrated that subnormal peak levels found in prepubertal obese children are not associated with higher growth velocity SDS, child height minus midparental height, and bone age when compared to their normal-weight peers.[16] On the other hand, states of undernutrition are characterized by variable degrees of GH resistance. Decreased total energy intake as well as isolated protein–calorie malnutrition and isolated nutrient deficiencies could result in elevated GH levels, but low levels of IGF1.[17]

2.2 Biochemical Aspects

A correct diagnosis is essential in the evaluation of patients with short stature for suspected GHD. A false-positive diagnosis will result in a prolonged, invasive, expensive, and unnecessary treatment. On the other hand, a false-negative result will lead to an unfavorable clinical outcome.

It is accepted that a unique cutoff value should be used for all the stimulation tests used for laboratory testing of GHD in children.[12] In general, the diagnosis is confirmed by two GH stimulation tests below a cutoff value, but they are currently more or less arbitrarily determined. In addition, inconsistencies within the results obtained using different analytical tools and laboratories, are interfering the reliability of this decision. There are several factors contributing more or less to the variability of GH results and thus, limiting the diagnostic power of the stimulation tests:

1. Factors modifying GH status, such as age, pubertal stage, nutritional status, possible abnormalities of other hormonal axes, etc.
2. Design of the GH immunoassay, preparation of the different assay calibrators, and matrix differences between them and samples, specificity of antibodies, and interference with endogenous human GH-binding proteins.

Recently, an iSYS assay (automated IDS-iSYS multidiscipline immunoassay system) has been developed.[18] This assay system fulfils current consensus determinations and could constitute a new tool for reliable measurement of GH. Moreover, this methodology could allow establishment of comparison of results with other GH commercially available assays. It is worth mentioning that iSYS assay has been compared with isotopic mass spectrometry (ID-MS). This latter methodology, recently introduced, constitutes an alternative to antibody-based methods and is exempted from interferences caused by nontarget proteins. At the same time, it provides the traceability of the International System of Units (SI) obviating the need to refer to a particular GH-RP. Thus, the expression of the cutoff values in terms of ID-MS GH concentrations make them independent of the different particular combination of antibody and RP used. For that reason, this procedure could be an important step allowing improvement the accuracy of the diagnosis of GHD in children and adolescents.[18]

Regarding IGF1 and IGFBP3, they increased with age in prepubertal children reaching the highest levels at puberty. Girls, in general, showed peak values approximately one year earlier than boys. Free IGF1 represents a small percentage of total IGF1 in the serum, with the majority forming a 150 kDa ternary complex with IGFBP3, and acid labile subunit (ALS). Some IGF1 also circulate in binary complexes with the various IGFBPs, and they could be measured when using immunoradiometric sandwich assay (IRMA). For this reason, to obtain an adequate measurement of IGF1, it is firstly necessary to separate the complexes, and this process can result in significant inter- and intraassay variability. Ethanol extraction is used to dissociate IGF1 from its binding proteins and precipitates large molecules from the sample, but IGFBP1 and IGF8P4 do not precipitate well and may affect results. Thus, other techniques as IGF2 saturation can be used to occupy the binding sites on the IGF binding proteins, and thereby increase free IGF1. Using this methodology jointly with ethanol precipitation, a further reduction of interference of binding proteins in IGF1 measurement can be obtained. Antibodies used for IGF1 can be polyclonal or monoclonal, where IGF2 saturation has been used. Cross-reactivity of the antibody with IGF2 must be very low or the results will be falsely elevated.[19]

The use of an automatic chemiluminescent assay system for IGF1 and IGFBP3, fulfilled the requirements as regards precision, specificity, linearity, and recovery. Results of reference values vary between published studies, due to methodological differences, different standards, and antibodies, age differences, etc. It is well-known that IGF1 levels are also influenced by other variables, such as nutritional status, severe trauma, physical training, some social factors, and upto 20% of GHD patients might present normal IGF1 values.[20] Moreover, it is important to take into account that regarding IGF1 and IGFBP3, reference ranges for young children (less than 2–3 years) are very controversial and not universally validated.

GH stimulation tests play a critical role in the diagnosis of GHD. It is worth mentioning that, in a revision of the literature, Sizonenko *et al.* showed that at least 34 provocative tests have been developed, and 189 different combination protocols were found.[7] Most of these tests lack adequate normative values for age, pubertal stage, and gender. Moreover, for ethical considerations, many of the normative data have been obtained in short children or children presenting with a decrease in growth velocity, that is, constitutionally short children with absence of endocrine disease or short children investigated for possible GHD.[7] Another issue to be taken into account is the relative poor intraindividual reproducibility, when the same tests are performed on different days. For that reason, Gundrud and Wilson proposed that stimulation tests should not be used for routine diagnosis of GHD in children, as these tests lack precision, accuracy and have little concordance with disease.[21] Moreover, these authors pointed out that stimulation tests are poor predictors of response to GH therapy, expensive, cumbersome, and in some of them, side effects are possible.[21] Some years ago, it was suggested that the administration of estrogens or testosterone a few days prior to a GH stimulation test could contribute to increase GH response, and distinguish between GHD and constitutional delay of growth and puberty.[22,23] Various protocols for priming have been used over the years with differing sex hormone preparations. Nevertheless, sex steroid priming is not currently employed by all pediatric endocrinologists.[23,24]

Another particular issue concerns GHD in the neonatal population. Random GH values within the first two weeks of life could be clinically useful, since the mean GH nadir is above 10–30 ng/mL.[25] GH levels should be obtained at the time of spontaneous hypoglycemia, and a "normal" level is likely around 30 ng/mL.[26] In this special group, IGFBP-3 could be a useful marker of GHD and IGFBP2 levels are high in GHD patients. GH stimulation tests in neonates are difficult to interpret based on the high baseline GH levels. Thus they would be unnecessary given the alternatives available.[21,27] IGF1 levels could be influenced by infant size, and may also be affected by sex, maternal smoking, ethnicity, and medication exposure. Breastfed infants have lower IGF1 levels, and this could be related to reduced weight gain.[27]

2.3 Imaging Aspects

Bone age assessment is an essential tool in the evaluation and follow-up of GHD patients, and it is important to take into account that those who have experienced prenatal disorders are at greater risk of developing a more important bone delay. GHD diagnosis prompts the imaging evaluation of the pituitary–hypothalamic region. Magnetic resonance imaging (MRI) has been considered as the method of choice for this evaluation. However, MRI results were normal in 87.5% of GHD cases. In patients with severe GHD, and in patients with multiple pituitary hormone deficiencies (MPHD), MRI is more likely to be abnormal.[28]

2.4 Molecular–Genetic Aspects

During the last few years, our knowledge of the development of the anterior pituitary gland have witnessed a significant advance as regards the process, by which sequential, temporal, and spatial expression of early and late transcription factors (HESX1, SOX2, SOX3, LHX3, LHX4, Ptx1, Ptx2 and PROP1, and POU1F1 among the most studied) and signaling molecules (FGF, BMP, Wnt, and SHH) results in shaping of the gland, cell linage, and commitment and hormone producing cells, including somatotrophic cells.[29] In this context, our understanding of genetic etiology of GHD isolated or in combination with other pituitary hormone deficiencies is explained. In addition, all these studies highlighted the role of autocrine and paracrine factors, with its possible implications for the etiology, pathophysiology and disease progression of both congenital and acquired GHD. Thus, we can observe in recent years an increasing contribution from genetics in patients with congenital GHD. However this diagnosis remains a subject of considerable controversy with remarkable variability in clinical practice (Table 2).[11]

Table 2 Genetic Etiology of GHD.

Genes	Hormonal Phenotypes
HESX1 (septo-optic displasia)	Variable involvement of pituitary hormones
PROP1	GH, PRL, TSH, LH, and FSH deficiencies; variable ACTH deficiency
POU1F1	GH and PRL deficiencies; variable degree of TSH deficiency
LHX3	GH, TSH, LH, FSH, and PRL deficiencies
LHX4	GH, TSH, and ACTH deficiencies
SOX3	GH
GLI2	(Holoprosencephaly) hypopituitarism
FGF8	(Holoprosencephaly) hypopituitarism
GLI3	(Pallister-Hall syndrome) Hypopituitarism
OTX2	TSH, GH, LH, and FSH deficiencies; variable ACTH and PRL deficiency
FGFR1	Hypopituitarism
IGSF1	TSH deficiency; variable GH and PRL deficiency

In most of the cases of children diagnosed with ISS, their shortness might be associated with a combination of multiple gene variations as suggested by genome-wide association studies, but in some cases, short stature may be the result of a monogenic disorder. The purpose of genetic testing in short stature children is to identify monogenic causes of short stature due to uncommon genetic variants with an important clinical impact. Identification of polygenic short stature is not currently feasible; and it could not be clinically useful. Monogenic causes of short stature are underdiagnosed in the pediatric clinical practice and it would be very useful to identify them for various reasons:[11]

1. To provide the family with an accurate answer as to why the child is not growing normally and provide, then, an adequate genetic counseling.
2. The genetic disorder may alert the clinician for other possible comorbidities, eventually associated at present or at a near future.
3. To assure whether GH treatment is a real valuable resource or it should not be taken into account.

There are several well-known genetic conditions that impair growth before and/or after birth; however, testing every short child for all possible genes involved would be impossible from a practical point of view. Therefore, a selection procedure would be necessary. One of the most important candidate genes to be tested is the SHOX gene, where complete or partial mutations and deletions have been described. Rappold *et al.* proposed to perform genetic testing for SHOX in all patients with a clinical disorder of Leri–Weill syndrome, Madelung deformity, and in short children with a first-degree relative with such abnormality.[30] Nevertheless, SHOX defects have also been encountered in approximately 2–4% of children diagnosed with ISS without the typical anatomical abnormalities.[31] Therefore, a clinical scoring system for individuals without these highly suggestive signs has been proposed.[32] Even though this score could be of great help for pediatric endocrinologists, its predictive value has not been adequately validated in normal populations.[32] Other potential candidate genes involved in children with short stature are GH1, GHR, STAT5B, and IGFALS, but mutations of these genes are not so frequent.[33,34]

Concerning SGA children, potential candidate genes to be studied are IGF1, IFG2, and IGFR. The presence of severe intrauterine growth retardation, postnatal growth failure, microcephaly, mental retardation, sensorineural deafness, and dysmorphic features should led to investigate an homozygous deletion or mutation of IGF1, while middler phenotypes could be related to heterozygosity for an IGF1 deletion or mutation.[31] Methylation abnormalities for IGF2 have been reported in patients with Silver–Russell syndrome.[35]

Some anatomical alterations such as pituitary hypoplasia associated with ectopic neurohypophysis or callosus corpora agenesis should prompt to evaluate mutations of PROP1 or HESX1 genes respectively. In these mutations, it is important to take into account the possibility of a progressive impairment of pituitary function and the late development of hormonal deficiencies.

Although more information has become available in recent years, it is still uncertain, which clinical features would have genetic predictive power. For this reason, the Consensus Guidelines for GHD in children proposed the following criteria:[12]

1. Early onset of growth failure
2. Positive familial history and possible consanguinity
3. Height below 3 SDS from the mean
4. Extremely low GH response to provocative tests, with very low IGF1 and IGFBP3 levels

Anyway, at present, tests for genetic mutations are only available in some laboratories and they are very expensive for their routine use. Nevertheless, comparative genomic hybridization could be employed for a genome-wide assessment of duplications or deletions that may have clinical consequences.[11]

2.5 Therapeutic Aspects

As regards treatment, some questions about when, how, duration, results, and safety arise. Many publications and several guidelines have been published;[9,12,36,37] but the main doubts are: are all these papers, as presented, helpful to the pediatric clinician or endocrinologist? Moreover, not all the recommendations have the some power of evidence; thus, another question might be: what is the overall usefulness of consensus guidelines? We believe that the answer is clearly positive However, there are several general and specific features that are worth discussing.

Firstly, in general, it is recommended that GH should be administered in the evening. Do we have nowadays enough evidence to assure any particular time of day for injections? This places an undeniable burden on patients and their families (e.g., during vacations, children sleeping over at a friend's house). Secondly, the dose–response relationship in terms of height velocity has been demonstrated using the GH doses suggested by all the guidelines, but under certain circumstances, and in some specific pathologies, higher doses than those used for GHD patients are required. Different prediction models of growth responses have been proposed, but they are quite complex for general practitioners and some of them are currently being investigated because they need further evaluation.[29]

Regarding monitoring of GH therapy, the evaluation of changes in height and in height velocity are the most important parameters in clinical practice. These data should always be expressed in SD per year in order to have confident and comparable results. In recent years, monitoring of serum IGF1 has been used for ensuring compliance and safety, although IGF1 levels does not always correlate well with growth response.[9,38] Measurements of leptin, bone markers, and lipid profile among others have been proposed as early predictors of therapeutic response,[12] although these measurements are not currently used in clinical practice.

Nevertheless, the concept of individualizing and optimizing GH therapy for children is gaining credence and several groups have proposed approaches to implement this strategy.[39,40] It should also be considered that there are some factors influencing response to GH treatment; it is well known that treatment at the youngest possible age correlates with better results in terms of height and some metabolic and cardiovascular outcomes. In children entering puberty at an inadequate height, the use of GH in higher doses or adding GnRH agonist therapy is being more often evaluated and used, and in these patients, with MPHD in whom puberty does not occur spontaneously, GH therapy should be initiated at the appropriate time taking into account social and psychological aspects. Another issue to be considered is treatment during the transition period from puberty to adulthood. This period is, in general, a lengthy and complex process that varies depending, partly on the various geographical regions and ethnic groups.[5,6,41] During this period, adult final height, adult body composition, peak bone mass, and psychosocial adult balance is attained. It is worth mentioning that, although in a majority of patients, GHD is diagnosed and treated during childhood, there is a subset of patients who acquire GHD during the transition period.[13] It seems that GH replacement therapy should be considered regardless of whether the deficiency arose during childhood or the transition period considering mainly, in this latter period, lipid profile, bone mineral content, and cardiovascular aspects.[13]

Concerning the side effects, they are very rare and these include prepubertal gynecomastia, arthralgia, benign intracranial hypertension, and in the absence of other risk factors, there is no clear evidence that the risk of brain tumor recurrence, leukemia, slipped capital femoral epiphysis or diabetes is increased during long-term GH treatment. Recently, a report of the Santé adult GH Enfant study (SAGhE) coming from the French Agency of Health

Safety raised concern regarding the long term safety of GH therapy.[42] The study was a retrospective evaluation of follow-up in patients who were treated during childhood for idiopathic, isolated GH deficiency, neurosecretory GH dysfunction, idiopathic short stature, and SGA. When compared with the French reference population an increased the all-cause mortality (standardized mortality ratio (SMR = 1.33), was found. The study reported an increase mortality due to diseases of the circulatory system (SMR = 3.07), subarachnoid intracerebral hemorrhage (SMR = 6.66), and bone tumor (SMR = 5.00). The mean period between initiation of GH therapy and the time of census or death was approximately 17 years. But, in contrast, the studies performed in Netherlands, Sweden, and Belgium did not demonstrate any death from cancer or cerebrovascular diseases.[43] In the Hypopituitary Control and Complications Study (HypoCCS), a global post marketing surveillance study conducted by Eli Lilly indicates no increased risks of mortality or incidence of cancer, stroke, and myocardial infarction in patients who have received pediatric GH treatment.[44] Other studies coming from KIGS/KIMS (a database sponsored by Pfizer) showed similar results.[45] Thus, summarizing all these different and somehow contradictory aspects,[46] we should also consider that people are at higher risk as they get older, have a strong family history of cancer or the values of IGF1 they have reached and maintained during all the time they were treated are clearly high. On the other hand, the potential influence of environmental and behavior aspects remains poorly understood and human populations comprise many varieties of genotypes that respond in a different manner to the same kind of exposures.[29,47] The morbidity assessment, evaluated in a recent study, showed an increase of cancer incidence, and the risk for second neoplasms in patients treated with GH after a primary cancer was significantly increased.[48] Nevertheless, in this paper, authors stated that these results may be affected by several biases and confounders which need to be addressed in further large-scale controlled surveys.[48]

Another important issue to be considered is treatment cost, which is given special attention by regulatory bodies of the Public Health System and insurance companies. It is well known that GH treatment has been demonstrated to be cost-effective and for this reason, in many countries, it is reimbursed by the Public Health Service System. Thus, inappropriate prescription (e.g., too much, too little, indications not yet approved, nonresponders), and/or inefficiency of some devices available or very difficult to be used by patients or their families could result in an unpredictable increase in the final cost with an evident impact on the health care system budget.[49]

3. OVERGROWTH AND GH EXCESS

Excessive growth is less common than undergrowth, and therefore, parents rarely seek medical advice unless other signs or symptoms are associated with tall stature. It is estimated that approximately 2.5% of the population has a height above 2 SDS.[15]

As we have already mentioned, longitudinal growth results from multifactorial and complex processes that take place in the context of different genetic traits and environmental influences.

Manifestations of overgrowth are variable. It may affect linear growth "in toto," resulting in tall adult height, or it may affect only one region (hemihypertrophy, macrocephaly, macrodactylia, etc).[50]

Familial tall stature, also known as "constitutional tall stature", is the most common cause of tall stature. The second most common cause of tall stature is nutritional. Hormonal causes of tall stature include hyperthyroidism, GH excess, and precocious puberty. Although in the latter, accelerated height velocity could be transient and final adult height is compromised due to premature epiphyseal fusion.

It is also worth mentioning that overgrowth syndromes comprise a heterogeneous group of disorders including excessive somatic and visceral growth. Many syndromes are characterized by substantial localized or asymmetric tissue overgrowth represented by Beckwith–Wiedemann syndrome, Sotos syndrome, Madelung's disease, etc. There are some important chromosomal causes as Klinefelter syndrome and autosomal recessive inborn error of aminoacid metabolism, such as homocystinuria due to cystathionine synthase deficiency (Table 3).[51]

Table 3 Overgrowth Syndromes with Typical Clinical Features.

- Klinefelter syndrome
- Marfan's syndrome
- Sotos syndrome
- Homocystinuria
- Beckwith–Wiedemann syndrome
- Weaver's syndrome
- Marshall–Smith syndrome
- Simpson–Golabi–Behmel syndrome

In conclusion, complex mechanisms involved in the regulation of somatic growth remain to be fully understood particularly at the molecular level of genetic field. Among various conditions showing overgrowth, genetic overgrowth syndromes, refer to nonhormonally mediated overgrowth conditions that can accompany increased height and/or head circumference, various degrees of mental retardation or physical dimorphism in children. There are overlaps in clinical and molecular features among overgrowth syndromes, which constitute a real burden for an accurate diagnosis.[50]

3.1 Diagnostic Aspects

Diagnosis constitutes an example of the multifactorial complexity involved and of the broad clinical spectrum of overgrowth. The following diagnostic tools should be taken into account:

- Medical history: tall constitutional and/or familial height? Are bone age and height prediction really helpful to define these clinical situations or—more likely—are they only able to provide relative guidance?
- Biochemical evaluation including GH/IGF axis assessment and evaluation of potential congenital errors of metabolism in patients with suggestive clinical symptoms.
- Standard genetic evaluation to rule out particularly trisomy X in women and men, with XXY or XYY, and fragile X syndrome.
- Imaging studies including the hypothalamic–pituitary region, long bones, and other specific studies depending on the suspected disease (for example, echocardiogram with aortic root measurement in Marfan's syndrome).

However, a real difficulty arises in patients with tall stature, no apparent symptoms of a specific syndrome, and tall parents. Are we in the presence of genetic tall stature, aromatase deficiency, the onset of pituitary gigantism, androgen resistance, among others? If the patient instead had mental retardation and certain dysmorphic features not clearly suggestive of a specific syndrome, with a normal karyotype, we should resort to high resolution comparative genomic hybridization array-based (HR-CGH) and multiplex ligation probe amplification, to detect possible microdeletions, and/or duplications, and potential candidate genes for specific phenotypes.[52]

3.2 Molecular–Genetic Aspects

The molecular causes of a handful of overgrowth syndromes have been partially elucidated, but despite many recent advances, the disease-causing mechanism remains unknown in a great number of patients (almost 20–40%), underlying the genetic heterogeneity of these disorders.[52]

Moreover, a large number of patients do not "belong" to any recognizable condition, remaining still unexplained. Several chromosomal duplications and deletions have been characterized in patients with overgrowth, suggesting that some unclassified syndromes may be caused by subtle genomic imbalance rearrangements (i.e., various types of NSD1 aberrations have been identified in patients with Sotos syndrome, including 5q35 microdelection; trisomy of 15q26qter is frequently associated with tall stature and mental retardation caused by duplication of the IGF1 receptor gene).[15,53,54]

These data strongly support the hypothesis that studying chromosomal aberration in overgrowth syndrome and using adequate sensitive tools may lead to the identification of an important number of cryptic chromosomal anomalies.

Concerning all these concepts, the benefits of a whole-genome array-CGH (comparative genomic hybridization) for the diagnosis of overgrowth syndromes should be highlighted. Besides, careful examination of patients may allow identification of novel clinical recognizable overgrowth conditions. Moreover, a detailed analysis of the rearranged regions may pave the way for the identification of genes involved in growth regulation. Perhaps, this could contribute to gain more accurate knowledge in view of the need for a specific tumor surveillance program, since some tumors are frequently associated with some overgrowth syndromes.[50,55]

Regarding obesity and tall stature, obese children tend to have increased growth velocity, although they rarely reach excessive adult height. Though rare, monogenic abnormalities have been reported in obese patients due to mutations of the leptin gene, leptin receptor, and also melanocortin-4 receptor (MC4R).[56] Tall stature in obese children would be mediated by hyperinsulinism, elevated IGF1 levels, particularly the free fraction, and low IGFBP1 levels, despite suppressed GH levels.

Another issue to be taken into account is the fact that skeletal maturation may have a role in the association between height and adiposity. Nevertheless, some studies have demonstrated that the association between height and adiposity in prepubertal children becomes weaker when adjusting by skeletal maturation. Furthermore, increased BMI in children on a path to becoming overweight adults precedes an advancement in skeletal development and subsequently tall stature during puberty.[57] The true burden lies in the fact that awareness of new monogenic conditions, associated with tall stature and obesity has substantially enlarged our knowledge. Thus, a different clinical management is required other than the "classical" approach currently used. A current issue to be considered is prenatal overgrowth.

Maternal diabetes is one of the most common causes, which appears to be related to the degree of maternal hyperglycemia and fetal hyperinsulinemia. The classical Beckwith–Weidemann syndrome or some of its variants with overgrowth and no malformations is a diagnosis that should be taken into account. Diagnostic burden at this stage constitutes a real difficulty, as single or multiple pancreatic islet β cells adenoma or some of the various forms of congenital hyperinsulinism should be considered.[15,58] Perlman syndrome, though rare, is characterized by prenatal overgrowth, renal dysplasia, viscer-omegaly. Therefore, differential diagnosis from Beckwith–Weidemann would be mandatory.[15]

3.3 Imaging Aspects

Some overgrowth syndromes (e.g., Proteus syndrome, Klippel–Trenaunay–Weber syndrome, Madelung's disease, and neurofibromatosis type I) have overlapping clinical manifestations, and in many cases, no established genetic tests for their diagnosis. In such cases, radiological methods are important contributory tools for the diagnosis of these syndromes.[50] Moreover, in some cases of neurofibromatosis type I, the presence of tumors of the peripheral nervous system, particularly plexiform neurofibromas, is confirmed by radiological examination. Tomographic findings depend on the histological features of the tumors and may exhibit soft tissue density.[59]

3.4 Therapeutic Aspects

Human overgrowth poses a real therapeutic challenge because of the great phenotypic variability and a large number of clinical symptoms and syndromes characterized by tall stature. Overall, treatment depends on diagnosis, evaluation of the extent and degree of involvement, prevention of potential secondary manifestations, quality of life, and in some cases, the prognosis of survival. Depending on etiologies, multidisciplinary teams integrated by pediatricians, geneticists, imaging specialists, orthopedists, cardiologists, ophthalmologists, and psychiatrists could be advisable.

Familial-tall stature or constitutional-tall stature is the most common cause of tall stature.[60] In the case of girls, who more often seek medical advice, treatment consists of the administration of high dose estrogens.[61] Results depend, in general, on the dose used and bone age at the initiation of treatment.[15] Given the severe side effects reported, namely thrombosis and a potential impairment of fertility,[62,63] these treatments should be indicated in selected cases under strict clinical monitoring. In boys with constitutional tall stature and predicted adult height well above the 97th percentile, long-acting testosterone esters have been used. Results, as in girls, are dose- and bone-age dependent and these agents may also have undesirable side effects.[60]

Another common cause of tall stature is nutrition. Height and weight are at higher percentiles, and bone age is moderately advanced. So, final adult height is not significantly compromised.[51,60] In these cases, only changes in diet and lifestyle are required.

Hormonal causes of tall stature include hyperthyroidism, precocious puberty, and GH excess. Hyperthyroidism is more common in girls, here bone age could be moderately advanced so final adult height is usually compromised. Treatment in general, consists of the administration of methimazole, carbimazole, or in certain circumstances radioiodine. In precocious puberty, although the child is tall at the onset, the ultimate adult height is compromised due to premature epiphyseal fusion. GH hormone excess causes gigantism in childhood. It is very rare and usually caused by GH-secreting pituitary adenoma (somatotropinoma) or hyperplasia of the somatotrophs. Associations include MEN-1 syndrome, Carney complex, and McCune Albright syndrome. Up to 20% of published cases of gigantism are parts of McCune Albright's syndrome.[15]

In patients with gigantism due to pituitary somatotropic adenoma, surgery,[64,65] and/or somatostatin analogs are the treatment of choice.[65–68] Transsphenoidal surgery is limited below the age of 8-9 years, when sphenoid sinus pneumatization has not been completed yet. Treatment with Pegvisomant, an antagonist of the GH receptor, could be used, still enough experience is not there in pediatric patients.[69] Radiotherapy, either primary or postsurgical, has a slow onset of treatment effect and a high treatment-related morbidity of panhypopituitarism.[65]

Regarding overgrowth syndromes of nonendocrine etiology, in some cases, sex steroids can be used with variable results and further experience is needed.[15] These agents are contraindicated in patients with homocystinuria[70] or other conditions associated with potential risks of thrombosis.[15,60]

Some of these syndromes may be associated with various types of tumors. Therefore, strict monitoring is essential during patient follow-up. Finally, recent knowledge of monogenic disorders associated with obesity and tall stature, constitutes a new therapeutic challenge.[56,71]

In conclusion, in the presence of clinical symptoms of overgrowth, an adequate clinical, genetic, and molecular diagnosis should be performed, as not only the possible and most appropriate therapeutic management should be evaluated, but also familial genetic counseling, when applicable.

REFERENCES

1. Dauber A, Yu Y, Turchin MC, et al. Genome-wide association of copy-number variation reveals an association between short stature and the presence of low-frequency genomic deletions. *Am J Hum Genet.* 2011;89:751–759.
2. Lango Allen H, Estrada K, Lettre G, et al. Hundreds of variants clustered in genomic loci and biological pathways affect human height. *Nature.* 2010;467:832–838.
3. Chan Y, Holmen OL, Dauber A, et al. Common variants show predicted polygenic effects on height in the tails of the distribution, except in extremely short individuals. *PLoS Genet.* 2011;7(12):e1002439.
4. Waldman LA, Chia DJ. Towards identification of molecular mechanisms of short stature. *Int J Pediatr Endocrinol.* 2013;2013(1):19.
5. Rosenfeld RG, Nicodemus BC. The transition from adolescence to adult life: physiology of the "transition" phase and its evolutionary basis. *Horm Res.* 2003;60:74–77.
6. Hochberg Z. Evo-devo of child growth II: human life history and transition between its phases. *Eur J Endocrinol.* 2009;160:135–141.
7. Sizonenko PC, Clayton PE, Cohen P, Hintz RL, Tanaka T, Laron Z. Diagnosis and management of growth hormone deficiency in childhood and adolescence. Part 1: diagnosis of growth hormone deficiency. *Growth Horm IGF Res.* 2001;11:137–165.
8. WHO Child Growth Standards. http://www.who.int/childgrowth/en/
9. Cohen P, Rogol AD, Deal CL, et al. Consensus statement on the diagnosis and treatment of children with idiopathic short stature: a summary of the Growth Hormone Research Society, the Lawson Wilkins Pediatric Endocrine Society, and the European Society for Paediatric Endocrinology Workshop. *J Clin Endocrinol Metab.* 2008;93:4210–4217.
10. Wit JM, Clayton PE, Rogol AD, Savage MO, Saenger PH, Cohen P. Idiopathic short stature: definition, epidemiology, and diagnostic evaluation. *Growth Horm IGF Res.* 2008;18:89–110.
11. Dauber A, Rosenfeld RG, Hirschhorn JN. Genetic evaluation of short stature. *J Clin Endocrinol Metab.* 2014;99:3080–3092.
12. Growth Hormone Research Society.Consensus guidelines for the diagnosis and treatment of growth hormone (GH) deficiency in childhood and adolescence: summary statement of the GH Research Society. *J Clin Endocrinol Metab.* 2000;85: 3990–3993.
13. Fideleff HL, Jonsson B, Koltowska-Häggström M, Boguszewski MC, Wilton P, Boquete HR. GH deficiency during the transition period: clinical characteristics before and after GH replacement therapy in two different subgroups of patients. *J Pediatr Endocrinol Metab.* 2012;25:97–105.
14. Van Dommelen P, de Kroon ML, Cameron N, Schönbeck Y, Van Buuren S. The impact of height during childhood on the national prevalence rates of overweight. *PLoS ONE.* 2014;9(1):e85769.
15. Argente J, Sotos JF. Overgrowth with and without obesity: clinical and molecular principles. *Ann Pediatr (Barcelona).* 2012;76(3):161e1–1628e.
16. Barrett J, Maranda L, Nwosu BU. The relationship between subnormal peak-stimulated growth hormone levels and auxological characteristics in obese children. *Front Endocrinol (Lausanne).* 2014;5:35.
17. Møller N, Jørgensen JO. Effects of growth hormone on glucose, lipid, and protein metabolism in human subjects. *Endocr Rev.* 2009;30:152–177.
18. Wagner IV, Paetzold C, Gausche R, et al. Clinical evidence-based cutoff limits for GH stimulation tests in children with a backup of results with reference to mass spectrometry. *Eur J Endocrinol.* 2014;171:389-3897.
19. Hawkes CP, Grimberg A. Measuring growth hormone and insulin-like growth factor-I in infants: what is normal? *Pediatr Endocrinol Rev.* 2013;11:126–146.

20. Rosenfeld RG, Albertsson-Wikland K, Cassorla F, et al. Diagnostic controversy: the diagnosis of childhood growth hormone deficiency revisited. *J Clin Endocrinol Metab.* 1995;80:1532–1540.
21. Gundrud LM, Wilson DM. Is growth hormone stimulation testing in children still appropriate? *Growth Horm IGF Res.* 2004;14:185–194.
22. Marin G, Domené HM, Barnes KM, Blackwell BJ, Cassorla FG, Cutler Jr GB. The effects of estrogen priming and puberty on the growth hormone response to standardized treadmill exercise and arginine-insulin in normal girls and boys. *J Clin Endocrinol Metab.* 1994;79:537–541.
23. Lanes R. Sex hormone priming. *J Pediatr Endocrinol Metab.* 2011;24:7–8.
24. Rosenbloom AL. Sex hormone priming for growth hormone stimulation testing in pre- and early adolescent children is evidence based. *Horm Res Paediatr.* 2011;75:78–80.
25. Pagani S, Chaler EA, Radetti G, et al. Variations in biological and immunological activity of growth hormone during the neonatal period. *Horm Res.* 2007;68:145–149.
26. Bozzola M, Tettoni K, Locatelli F, et al. Postnatal variations of growth hormone bioactivity and of growth hormone-dependent factors. *Arch Pediatr Adolesc Med.* 1996;150: 1068–1071.
27. Hawkes CP, Grimberg A. Measuring growth hormone and insulin-like growth factor-I in infants: what is normal? *Pediatr Endocrinol Rev.* 2013;11:126–146.
28. Naderi F, Eslami SR, Mirak SA, et al. Effect of growth hormone deficiency on brain MRI findings among children with growth restrictions. *J Pediatr Endocrinol Metab.* 2015;28: 117–123.
29. Alatzoglou KS, Webb EA, Le Tissier P, Dattani MT. Isolated growth hormone deficiency (GHD) in childhood and adolescence: recent advances. *Endocr Rev.* 2014;35:376–432.
30. Rappold G, Blum WF, Shavrikova EP, et al. Genotypes and phenotypes in children with short stature: clinical indicators of SHOX haploinsufficiency. *J Med Genet.* 2007;44: 306–313.
31. Caliebe J, Broekman S, Boogaard M, et al. IGF1, IGF1R and SHOX mutation analysis in short children born small for gestational age and short children with normal birth size (idiopathic short stature). *Horm Res Paediatr.* 2012;77:250–260.
32. Rappold GA, Fukami M, Niesler B, et al. Deletions of the homeobox gene SHOX (short stature homeobox) are an important cause of growth failure in children with short stature. *J Clin Endocrinol Metab.* 2002;87:1402–1406.
33. Walenkamp MJ, Wit JM. Genetic disorders in the growth hormone–insulin-like growth factor-I axis. *Horm Res.* 2006;66:221–230.
34. Rosenfeld RG, Hwa V. The growth hormone cascade and its role in mammalian growth. *Horm Res.* 2009;71:36–40.
35. Azzi S, Rossignol S, Steunou V, et al. Multilocus methylation analysis in a large cohort of 11p15-related foetal growth disorders (Russell Silver and Beckwith Wiedemann syndromes) reveals simultaneous loss of methylation at paternal and maternal imprinted loci. *Hum Mol Genet.* 2009;18:4724–4733.
36. Frasier SD. Editorial: the diagnosis and treatment of childhood and adolescent growth hormone deficiency–consensus or confusion? *J Clin Endocrinol Metab.* 2000;85:3988–3989.
37. Cook DM, Rose SR. A review of guidelines for use of growth hormone in pediatric and transition patients. *Pituitary.* 2012;15:301–310.
38. Cohen P, Germak J, Rogol AD, Weng W, Kappelgaard AM, Rosenfeld RG. American Norditropin Study Group.Variable degree of growth hormone (GH) and insulin-like growth factor (IGF) sensitivity in children with idiopathic short stature compared with GH-deficient patients: evidence from an IGF-based dosing study of short children. *J Clin Endocrinol Metab.* 2010;95:2089–2098.
39. Geffner ME, Dunger DB. Future directions: growth prediction models. *Horm Res.* 2007;68: 51–56.

40. Kriström B, Aronson AS, Dahlgren J, et al. Growth hormone (GH) dosing during catch-up growth guided by individual responsiveness decreases growth response variability in prepubertal children with GH deficiency or idiopathic short stature. *J Clin Endocrinol Metab.* 2009;94:483–490.
41. Lanes R, Boguszewski CL, Calzada R, Cassorla F, Fideleff H, Boquete H. Growth hormone deficiency: transition from adolescence to adulthood. *J Pediatr Endocrinol Metab.* 2010;23:225–233.
42. Carel JC, Ecosse E, Landier F, et al. Long-term mortality after recombinant growth hormone treatment for isolated growth hormone deficiency or childhood short stature: preliminary report of the French SAGhE study. *J Clin Endocrinol Metab.* 2012;97:416–425.
43. Sävendahl L, Maes M, Albertsson-Wikland K, et al. Long-term mortality and causes of death in isolated GHD, ISS, and SGA patients treated with recombinant growth hormone during childhood in Belgium, The Netherlands, and Sweden: preliminary report of 3 countries participating in the EU SAGhE study. *J Clin Endocrinol Metab.* 2012;97:213–217.
44. Mo D, Hardin DS, Erfurth EM, Melmed S. Adult mortality or morbidity is not increased in childhood-onset growth hormone deficient patients who received pediatric GH treatment: an analysis of the Hypopituitary Control and Complications Study (HypoCCS). *Pituitary.* 2014;17:477–485.
45. Wilton P, Mattsson AF, Darendeliler F. Growth hormone treatment in children is not associated with an increase in the incidence of cancer: experience from KIGS (Pfizer International Growth Database). *J Pediatr.* 2010;157:265–270.
46. Rosenfeld RG, Cohen P, Robison LL, et al. Long-term surveillance of growth hormone therapy. *J Clin Endocrinol Metab.* 2012;97:68–72.
47. Fudvoye J, Bourguignon JP, Parent AS. Endocrine-disrupting chemicals and human growth and maturation: a focus on early critical windows of exposure. *Vitam Horm.* 2014;94:1–25.
48. Deodati A, Ferroli BB, Cianfarani S. Association between growth hormone therapy and mortality, cancer and cardiovascular risk: systematic review and meta-analysis. *Growth Horm IGF Res.* 2014;24:105–111.
49. Spandonaro F, Cappa M, Castello R, Chiarelli F, Ghigo E, Mancusi L. The impact of real practice inappropriateness and devices' inefficiency to variability in growth hormone consumption. *J Endocrinol Invest.* 2014;37:979–990.
50. Lacerda LS, Alves UD, Zanier JF, Machado DC, Camilo GB, Lopes AJ. Differential diagnoses of overgrowth syndromes: the most important clinical and radiological disease manifestations. *Radiol Res Pract.* 2014;2014:947451 10.1155/2014/947451.
51. Kumar S. Tall stature in children: differential diagnosis and management. *In J Pediatr Endocrinol.* 2013;P53.
52. Malan V, Chevallier S, Soler G, et al. Array-based comparative genomic hybridization identifies a high frequency of copy number variations in patients with syndromic overgrowth. *Eur J Hum Genet.* 2010;18:227–232.
53. Kurotaki N, Imaizumi K, Harada N, et al. Haploinsufficiency of NSD1 causes Sotos syndrome. *Nat Genet.* 2002;30:365–366.
54. Chen CP, Lin SP, Lin CC, et al. Molecular cytogenetic analysis of de novo dup(5)(q35.2q35.3) and review of the literature of pure partial trisomy 5q. *Am J Med Genet A.* 2006;140:1594–1600.
55. Neylon OM, Werther GA, Sabin MA. Overgrowth syndromes. *Curr Opin Pediatr.* 2012;24:505–511.
56. Martos-Moreno GA, Argente J. Paediatric obesities: from childhood to adolescence. *An Pediatr (Barcelona).* 2011;75:63e1–723e.
57. Van Dommelen P, de Kroon ML, Cameron N, Schönbeck Y, van Buuren S. The impact of height during childhood on the national prevalence rates of overweight. *PLoS ONE.* 2014;9(1):e85769.

58. De León DD, Stanley CA. Mechanisms of Disease: advances in diagnosis and treatment of hyperinsulinism in neonates. *Nat Clin Pract Endocrinol Metab.* 2007;3:57–68.
59. Fortman BJ, Kuszyk BS, Urban BA, Fishman EK. Neurofibromatosis type 1: a diagnostic mimicker at CT. *Radiographics.* 2001;21:601–612.
60. Davies JH, Cheetham T. Investigation and management of tall stature. *Arch Dis Child.* 2014;99:772–777.
61. Radivojevic U, Thibaud E, Samara-Boustani D, Duflos C, Polak M. Effects of growth reduction therapy using high-dose 17 beta-estradiol in 26 constitutionally tall girls. *Clin Endocrinol (Oxford).* 2006;64:423–428.
62. Venn A, Bruinsma F, Werther G, et al. Oestrogen treatment to reduce the adult height of tall girls: long-term effects on fertility. *Lancet.* 2004;29:1513–1518.
63. Hendriks AE, Drop SL, Laven JS, Boot AM. Fertility of tall girls treated with high-dose estrogen, a dose-response relationship. *J Clin Endocrinol Metab.* 2012;97:3107–3114.
64. Abe T, Tara LA, Ludecke DK. Growth hormone-secreting pituitary adenomas in childhood and adolescence: features and results of transnasal surgery. *Neurosurgery.* 1999;45: 1–10.
65. Keil MF, Stratakis CA. Advances in the diagnosis, treatment, and molecular genetics of pituitary adenomas in childhood. *US Endocrinol.* 2009;1:81–85.
66. Eugster EA, Pescovitz O. Gigantism. *J Clin Endocrinol Metab.* 1999;84:4379–4384.
67. Colao A, Pivonello R, Di Somma C, Tauchmanovà L, Savastano S, Lombardi G. Growth hormone excess with onset in adolescence: clinical appearance and long-term treatment outcome. *Clin Endocrinol (Oxford).* 2007;66:714–722.
68. Shimatsu A, Teramoto A, Hizuka N, Kitai K, Ramis J, Chihara K. Efficacy, safety, and pharmacokinetics of sustained-release lanreotide (lanreotide autogel) in Japanese patients with acromegaly or pituitary gigantism. *Endocr J.* 2013;60:651–663.
69. Rix M, Laurberg P, Hoejberg AS, Brock-Jacobsen B. Pegvisomant therapy in pituitary gigantism: successful treatment in a 12-year-old girl. *Eur J Endocrinol.* 2005;153:195–201.
70. Hernández M, Argente J. Hipercrecimientos. In: Cruz M, ed. *Tratado Pediatr.* Majadahonda, Madrid: Editorial Ergon; 2011; 10 (1): 947–955
71. Sabin MA, Werther GA, Kiess W. Genetics of obesity and overgrowth syndromes. *Best Pract Res Clin Endocrinol Metab.* 2011;25:207–220.

CHAPTER EIGHT

GH and Pituitary Hormone Alterations After Traumatic Brain Injury

Züleyha Karaca, Fatih Tanrıverdi, Kürşad Ünlühızarcı, Fahrettin Kelestimur[1]
Department of Endocrinology, Erciyes University Medical School, Kayseri, Turkey
[1] Corresponding author: e-mail address: fktimur@erciyes.edu.tr

Contents

Abstract

Traumatic brain injury (TBI) is a crucially important public health problem around the world, which gives rise to increased mortality and is the leading cause of physical and psychological disability in young adults, in particular. Pituitary dysfunction due to TBI was first described 95 years ago. However, until recently, only a few papers have been published in the literature and for this reason, TBI-induced hypopituitarism has been neglected for a long time. Recent studies have revealed that TBI is one of the leading causes of hypopituitarism. TBI which causes hypopituitarism may be characterized by a single head injury such as from a traffic accident or by chronic repetitive head trauma as seen in combative sports including boxing, kickboxing, and football. Vascular damage, hypoxic insult, direct trauma, genetic predisposition, autoimmunity, and neuroinflammatory changes may have a role in the development of hypopituitarism after TBI. Because of the exceptional structure of the hypothalamo-pituitary vasculature and the special anatomic location of anterior pituitary cells, GH is the most commonly lost

Progress in Molecular Biology and Translational Science, Volume 138
ISSN 1877-1173
http://dx.doi.org/10.1016/bs.pmbts.2015.10.010

hormone after TBI, and the frequency of isolated GHD is considerably high. TBI-induced pituitary dysfunction remains undiagnosed and therefore untreated in most patients because of the nonspecific and subtle clinical manifestations of hypopituitarism. Treatment of TBI-induced hypopituitarism depends on the deficient anterior pituitary hormones. GH replacement therapy has some beneficial effects on metabolic parameters and neurocognitive dysfunction. Patients with TBI without neuroendocrine changes and those with TBI-induced hypopituitarism share the same clinical manifestations, such as attention deficits, impulsion impairment, depression, sleep abnormalities, and cognitive disorders. For this reason, TBI-induced hypopituitarism may be neglected in TBI victims and it would be expected that underlying hypopituitarism would aggravate the clinical picture of TBI itself. Therefore, the diagnosis and treatment of unrecognized hypopituitarism due to TBI are very important not only to decrease morbidity and mortality due to hypopituitarism but also to alleviate the chronic sequelae caused by TBI.

1. INTRODUCTION

Traumatic brain injury (TBI) is a significant public health problem worldwide and results in long-term neurologic sequelae and increased mortality. It has been reported that the overall annual incidence of TBI is about 506.4 per 100,000 population and 235,000 victims are hospitalized each year because of nonfatal TBI in the United States.[1] The mean incidence rate of hospitalized and fatal TBI in Europe has been reported as 235 per 100,000.[2] A high rate of TBI annual incidence and mortality due to TBI in the rest of the world including Australia as 322 per 100,000 and New Zealand as ranging from 1100 per 100,000 to 2300 per 100,000 has also been reported.[3,4] The causes of TBI are road traffic accidents, which account for 50% of all cases, falls, violence-related incidents, work injuries, sports-related head injuries, and war accidents.[5,6] The increasing number of cars on the road and the increased rate of alcohol consumption among the young generation in particular are associated with more frequent traffic accidents and therefore more common sequelae due to TBI. Cyran reported hypopituitarism in TBI for the first time about 95 years ago.[7] Until recently, anterior pituitary dysfunction after TBI had not been given enough attention and was recognized as a rare endocrine abnormality. Benvenga *et al.* reported 367 cases of hypopituitarism due to TBI in 2000.[8] That study stimulated the endocrine community and a number of studies from different parts of the world were subsequently published in the literature confirming the obvious relation between TBI and neuroendocrine changes, mainly anterior pituitary insufficiency. Sports-related head injury including single trauma or chronic repetitive head trauma may also give rise to pituitary dysfunction.[9] War accidents which include falls, motor vehicle accidents, shrapnel, bullet wounds, and blastic brain injuries are causes of neuroendocrine abnormalities and have recently been paid more attention.[10] In this chapter, the current data regarding anterior pituitary dysfunction in adults after TBI due to traffic accidents, falls, sports activities, and war accidents are discussed.

2. NEUROENDOCRINE CHANGES IN THE ACUTE PHASE OF TBI

The acute phase is generally defined as the first 10–14 days after TBI. Hormonal changes which occur during the acute phase have been investigated in order to understand whether the changes are due to a critical illness or are TBI-specific, and to determine whether treatment is required and they have some predictive value? Tanriverdi *et al.* investigated the pituitary functions in the early acute phase and the relationships among hormone levels, severity of trauma, and mortality due to trauma, within 24 h of trauma, in 104 patients with TBI. Gonadotropins were the most commonly deficient anterior pituitary hormones in the early acute phase of TBI. Changes in cortisol, prolactin, and total testosterone levels were related to the severity of trauma as measured by the Glasgow coma scale (GCS), while no relation was found between hormonal changes and mortality. Age and GCS were significantly related to mortality.[11] In another study in which the patients underwent basal hormonal evaluation within 24 h of admission to the intensive care unit, 12% of the patients had hyperprolactinemia, 51.9% had low T3 syndrome, 5.8% had TSH deficiency, 41.6% had gonadotropin deficiency, 9.8% had ACTH deficiency, and 20.4% had growth hormone deficiency (GHD). Pituitary functions improved or worsened in a considerable number of patients after 12 months. On the other hand, it was shown that IGF-I levels lower than 84 ng/mL in the acute phase had a predictive value for GHD at 12 months.[12] Early posttraumatic hormone changes (0–12 days) were not found to be predictive, but may be indicative in some patients. However, hypoadrenalism in the acute phase should be kept in mind because untreated hypoadrenalism may be life-threatening.[13] Prasanna *et al.* investigated hormonal changes in 100 patients who suffered from moderate to severe TBI. The basal hormone levels were measured within 24 h of admission and they were repeated 7 days after TBI in patients who survived. The most commonly deficient hormone was growth hormone (GH) on day 7 and the most commonly increased hormone was cortisol on day 1.[14] The authors also found that severe TBI, increased intracranial pressure, and low GCS were risk factors for the development of hypopituitarism. Hormonal analysis was suggested in patients with moderate to severe TBI.[14] In that study, the percentages of low T3 syndrome in the acute phase on days 1 and 7 after TBI were 22 and 10.6%, respectively. Hannon *et al.* assessed the pituitary functions on day 10 following TBI in 100 patients with TBI and they found that 78% of the patients had low plasma cortisol level and these patients were associated with increased mortality and long-term hypopituitarism. Central diabetes insipidus was also a risk factor for increased mortality and long-term pituitary dysfunction.[15] Therefore, it seems that most of the hormonal changes seen during the acute phase of TBI reflect the acute adaptive response to the traumatic event and are transient. Hormonal changes may be related to the severity of trauma; the more the changes seen in hormone levels, the more severe the TBI. Replacement therapy for TSH, GH, and gonadotropin deficiencies during the acute phase is not recommended. ACTH deficiency immediately after TBI should be treated urgently, because it is associated with increased risk of death. In the diagnosis of adrenal insufficiency, besides clinical picture compatible with hypoadrenalism, measurement of basal serum cortisol level is required. Cortisol concentration less than 15 μg/dL is generally accepted as a diagnostic criterion for adrenal insufficiency in critical illness.[16] For glucocorticoid replacement, hydrocortisone (i.v.) is preferred and the duration of glucocorticoid replacement depends on the clinical situation and assessment of the HPA axis after the patient's situation has improved.

3. FREQUENCY OF HYPOPITUITARISM AFTER TRAUMATIC BRAIN INJURY

Until 2000, TBI was recognized as a rare cause of hypopituitarism. In that year, Benvenga *et al.* reported the data of 367 patients with TBI, which were published in the literature in a review.[8] During the last 15 years some retrospective and prospective studies were launched in order to find out the real frequency of pituitary dysfunction following TBI. The current data suggests that hypopituitarism due to TBI is not uncommon, in contrast to what was suggested previously, and in some patients hypopituitarism was undiagnosed and so untreated.[17–19] The frequency of hypopituitarism varies hugely in different studies from 0.9% to 68.5%.[19–23] The prevalence of hypopituitarism and GH deficiencies are shown in Table 1.

Table 1 Prevalence of Hypopituitarism and GH Deficiency in TBI.

References	Number	Severity (GCS)	Time After Event	Test(s) Used	GHD (%)	Total Hypopituitarism (%)	Study Design
Lieberman *et al.* (2001)	70	32 of patients ≤8	1–15 months	GST	14.6	68.5	Cross-sectional
Agha *et al.* (2004)	102	3–13	6–36 months	GST (ITT/ GHRH + arginine)	12.7; 7.8		Cross-sectional
Bondanelli *et al.* (2004)	70	3–14	3–6–9 months	GHRH + arginine	16	21	Prospective
Popovic *et al.* (2004)	67	9–13	44 months (median)	GHRH + GHRP-6	15	34.3	Cross-sectional
Aimaretti *et al.* (2005)	70	(3–15) 33: mild (15–13) 22: mod (12–9) 15: severe (≤8)	3–12 months	GHRH + arginine	20	22.2	Prospective
Leal-Cerro *et al.* (2005)	99	≤8 (severe)	Over 5 years >12 months	GHRH + GHRP-6 ITT, GST	5.8	24.7	Cross-sectional
Herrmann *et al.* (2006)	76	≤8 (severe)	5–47 months	GHRH + arginine, ITT	8	24	Cross-sectional
Schneider *et al.* (2006)	78/70	3–15	3–12 months	IGF, GHRH + arginine	10	36	Prospective
Tanriverdi *et al.* (2006)	52	3–15	24 h – 12 months	GHRH + GHRP-6	37.2	50.9	Prospective

(Continued)

Table 1 Prevalence of Hypopituitarism and GH Deficiency in TBI.—cont'd.

References	Number	Severity (GCS)	Time After Event	Test(s) Used	GHD (%)	Total Hypopituitarism (%)	Study Design
Klose *et al.* (2007)	104	3–15	10–27 months	ITT, GHRH + arginine	15	16	Cross-sectional
Klose *et al.* (2007)	46	3–15	12 months	ITT, GHRH + arginine	11	11	Prospective
Bavisetty *et al.* (2008)	70	3–14	3–6–9 months	GHRH-arginine	16	21	Prospective
Tanriverdi *et al.* (2008)	30	3–15	3 years	GHRH + GHRP-6 GST	23.3	29.9	Prospective
Berg *et al.* (2010)	246	3–12	4–47 months	GHRH + arginine, ITT,	5	21	Cross-sectional
Kokshoorn *et al.* (2011)	112	3–15	1–12 years	GHRH + arginine + ITT	3.6	5.4	
Tanriverdi *et al.* (2013)	25	3–15	5 years	GHRH + GHRP-6 GST	28	32	Prospective
Ulfarsson *et al.* (2013)	51	3–8	2–10 years	GHRH + arginine	21.6	27.5	Cross-sectional
Nemes *et al.* (2014)	126	3–12	3.9–2.5 years	ITT, GST, arginine	39.7	57.1	Cross-sectional

In a study including 102 patients who survived moderate to severe TBI, 10.7% of the patients had GHD, whereas 7.8% had severe GHD according to both the provocative tests including GST plus ITT and GST plus GHRH + arginine tests. Patients were assessed at a median of 17 months (range 6–36 months) after TBI.[24] A multicenter study in Italy by Aimaretti *et al.* which included 70 patients with TBI (mild to moderate to severe) revealed that 20% of the patients had severe GHD according to the GHRH + arginine test, at 12 months after TBI. Peak GH response less than 9.0 μg/L was accepted as severe GHD. Of the patients, 22.2% had hypopituitarism.[25] In a German cross-sectional study, 76 patients with severe TBI were investigated. In that study, 24% of the patients had hypopituitarism and 8% had GHD when tested by basal IGF-I and GHRH + arginine test. ITT was used for confirmation of the results in 7 of 12 patients who had a peak GH response less than 9.0 μg/L after the GHRH + arginine test.[26] Of 99 patients with severe TBI, 44 who had IGF-I in the lower range or who had a pituitary hormone deficiency other than GH were further investigated by GHRH + GHRP-6, ITT, and GST in a Spanish study. Gonadotrophin deficiency was the most common pituitary deficit and 24.7% of the patients had hypopituitarism, and GHD was diagnosed in 6% of the patients.[27] In a prospective study conducted by Schneider *et al.*, 78 patients with TBI underwent testing for pituitary function at 3 and 12 months (70 patients) after the injury. Of the patients, 36% had hypopituitarism 12 months after TBI. The GH axis was assessed by basal IGF-I and GHRH + arginine tests. The diagnosis of GHD was given when peak GH response was less than 9 ng/mL (μg/L) after the GHRH + arginine test.[28] The same authors also studied the predictors of anterior pituitary insufficiency in patients with TBI. Computed tomography scans and, in equivocal cases, MRI were obtained. Diffuse axonal injury (DAI), basal skull fracture, and older age were found to be major risk factors of posttraumatic hypopituitarism.[29] Supporting these data, Jeong *et al.* found that approximately one-third of patients with DAI had hypopituitarism.[30] Tanriverdi *et al.* reported a prospective study in which 52 patients were assessed while in intensive care within 24 h and 12 months after TBI. At 12 months, 50.9% of the patients had at least one anterior pituitary hormone deficiency. In the investigation of GHD, the GHRH + GHRP-6 test was used and a peak GH level less than 10 μg/L was taken as the cut-off level. Of the patients, 32.7% had GHD. The authors also suggested that pituitary function may improve or worsen in a considerable number of patients 1 year after TBI.[12] Klose *et al.* assessed the prevalence of hypopituitarism following TBI. The patients were investigated in the early acute phase and at 12 months.

Most of the patients had mild TBI (mTBI; 22 of 46 patients, only 15 had severe TBI) and long-term hypopituitarism was more frequent in severe TBI. Of the patients, 11% developed hypopituitarism 1 year after TBI. All patients with hypopituitarism also had GHD diagnosed by basal IGF-I, ITT, and GHRH + arginine if ITT was contraindicated.[13] Klose *et al.* reported another study in 2007; this was a prospective study on 104 patients with TBI. The patients were assessed at 13 (10–27) months following injury. Hypopituitarism was diagnosed in 15% of the patients and GHD was the most common deficit; the frequency was 15%. Increased intracerebral pressure and increased severity were found to have a predictive value for hypopituitarism.[31] In a study from Germany, conducted on 55 patients with TBI in whom the GCS varied from 3 to 15, 24.5% of the patients had hypopituitarism. The GHRH stimulation test alone was used in the investigation of GHD and it was defined in only one patient.[32] Berg *et al.* reported that 21% of the patients with moderate to severe TBI had hypopituitarism, of whom 1% had total, 2% had partial, and 18% had isolated hypopituitarism in a cross-sectional study including 246 patients. The study was carried out under the German Multi-centre Screening Programme. Of the patients with pituitary insufficiency, 12% had hypothyroidism, 9% had hypogonadism, 5% had GHD, and 4% had hypocortisolism. In that study, the dynamic test was carried out in the investigation of GHD only if IGF-I was <-2 SDS. Therefore, the prevalence of GHD may be underestimated.[33] Ulfarsson *et al.* investigated 51 patients with severe TBI and showed that 27.5% had hypopituitarism and 21.6% had isolated GHD according to the GHRH + arginine stimulation test. Screening for hypopituitarism after TBI was recommended only for overweight patients.[34] Nemes *et al.* recently reported a prospective study in which 126 TBI survivors had moderate to severe injury. The prevalence of hypopituitarism was defined in 57.1% of the patients. GHD/GHI was found to be the most common hormonal abnormality (39.7%), followed by hypogonadism. Neurosurgical intervention has been found as an independent risk factor.[35] The prevalence of hypopituitarism after TBI has been found to be lower in childhood than that in adults.[36] In a review published 8 years ago, Schneider *et al.* identified 10 studies, including 809 patients with TBI, in which the patients were assessed at least 5 months after the trauma and reported the prevalence of any pituitary deficiency as 27.5% and multiple pituitary hormone deficiency as 7.7%. GH and gonadotropins were found as the most commonly lost hormones.[37] In another review by Kokshoorn published in 2010 including 14 studies, large variations in the prevalence of hypopituitarism were found between the studies and they emphasized the role of different endocrinologic tests and criteria as the underlying factors responsible for variable results.[38] Klose *et al.* recently reported a much lower prevalence of GHD: only 1% of patients had GHD. After ITT, 5% of the patients had GHD but this decreased to 1% when confirmed with the pyridostigmine-GHRH test, which is not used in routine practice and is not well standardized. Confirmation of GHD diagnosed after ITT, which is accepted as the gold standard test for the diagnosis of GHD, has not been assessed previously with the pyridostigmine-GHRH test. The most important limitation of that study is that most of the study population had mTBI.[39] Since a huge number of people are subjected to TBI around the world every day, even low rates of GHD mean a very high number of patients with TBI might have GHD. The frequency of hypopituitarism varies between the studies published in the literature so far. These variable results depend on the type (closed, blastic injury, direct) and severity of trauma (mild, moderate, and severe according to GCS), number of trauma, study population (e.g., athletes in whom the trauma is repetitive), the design of the study, the diagnostic tests and criteria of hormone deficiencies, the screening procedure, tests and cut-off values used, and presence of confirmatory test.

The natural history of hypopituitarism after TBI is not well known, because of insufficient prospective data. Tanriverdi *et al.* conducted a 5-year prospective study, in which the patients were seen in the acute phase, within 24 h, and at 1, 2, and 5 years after TBI. GH was found to be the most common anterior pituitary hormone deficit at 1, 3, and 5 years following TBI. Interestingly, most of the pituitary hormone deficiencies improved over the 5 years, but a substantial number of patients still had hypopituitarism at the 5th year. Although rarely, new-onset hypopituitarism appeared or pituitary dysfunction worsened over the years in some patients. Taking these findings into account in the screening of pituitary functions, at least 5 years of follow-up may be recommended in some subgroups.[12,40,41] The risk factors for hypopituitarism after TBI are shown in Table 2. Although controversy exists about some of them, all the risk factors listed in the table depend on the studies published in the literature.

4. SPORTS-RELATED BRAIN INJURY AND HYPOPITUITARISM

Until recently, pituitary dysfunction after sports-related brain injury has been completely unnoticed. Only some case reports have appeared in the literature. The data are still scanty, except for some studies on boxers,

Table 2 Risk Factors for Hypopituitarism After TBI.

1. Older age
2. Increased intracerebral pressure in ICU
3. Longer stay in ICU
4. Autoimmunity (APAs and AHAs)
5. Severity of TBI
6. Radiological abnormalities
   a. Skull-base fracture
   b. Diffuse axonal injury
   c. Diffuse brain swelling
7. Genetic predisposition
8. Neurosurgical intervention for head injury
9. Cardiorespiratory arrest and mechanical ventilation in ICU
10. Central diabetes insipidus

kickboxers, and football players. Sports-related brain injury may be acute or chronic. An athlete may suffer from a TBI which is similar to other causes of TBI such as a fall or traffic accident, in terms of underlying pathophysiology and consequences, and it may happen once or, rarely, a couple of times. TBI may occur in noncombative sports, including hockey, swimming, and snowboarding. It may also occur in combative sports and is characterized by chronic repetitive head trauma, such as in boxing and kickboxing in which an athlete may be subject to head trauma more than 1000 times.[42–45] Concussion is the typical injury associated with sports, especially in combative sports including boxing, martial arts, football, and ice-hockey. Boxing is accepted as a worldwide authorized and very common sporting activity despite heavy criticisms because of the neurologic abnormalities observed in boxers. Boxing is characterized by accumulated damage to the brain from multiple punches and most boxers experience memory disturbances. Chronic traumatic encephalopathy (CTE) results from repetitive head blows and causes neurologic abnormalities in boxers.[46,47] Of retired professional boxers, 20% develop CTE.[48] Kelestimur *et al.* described hypopituitarism in boxers for the first time in 2004. In that preliminary study, 11 actively competing or retired male boxers and BMI-matched healthy nonboxing controls were compared in terms of pituitary functions. In the investigation of pituitary functions, the basal hormone levels were measured and the GHRH + GHRP-6 stimulation test was used in the assessment of GH status. In 45% of the boxers, GHD was diagnosed. The IGF-I levels in boxers were significantly lower than in healthy subjects. There was a significant negative correlation between peak GH levels and boxing duration, and between peak GH levels and number of bouts.[42] Kickboxing is one of the most popular martial arts (fighting style and technique) and head injuries are very common in both amateur and professional kickboxers because the use of punching, kicking, and, under some rules, kneeing and elbowing is permitted.[49] Tanriverdi *et al.* investigated pituitary function in both retired and active amateur kickboxers. The IGF-I levels were found to be significantly lower in kickboxers when compared with healthy subjects and 22.7 and 9.1% of the kickboxers had GH and ACTH deficiency, respectively. The GH axis was assessed by GHRH + GHRP-6 and GST.[44] Acute head trauma due to kickboxing may cause transient pituitary dysfunction such as secondary hypogonadism and slightly elevated prolactin level.[50] Pituitary volume in retired boxers with GHD is smaller than that in GH-sufficient boxers[45] (Fig. 1). In order to understand the role of autoimmunity, which plays a role in noncombative brain injury, in the pituitary dysfunction due to chronic repetitive head trauma, antipituitary antibodies (APAs) and antihypothalamus antibodies (AHAs) were detected. The high AHAs titers were found to be associated with low GH reserve in retired boxers. In the light of this finding, it could be suggested that hypothalamic damage may be more important than pituitary damage in boxers.[51] Kelly *et al.* reported that 23.5% of retired football players with a relatively low quality of life had hypopituitarism.[43] Sports-related head injury, particularly in boxing, may result in cognitive dysfunction in active or retired athletes.[52] Therefore, we investigated the relation between GH axis and cognitive performance in boxers and kickboxers. The minimental state

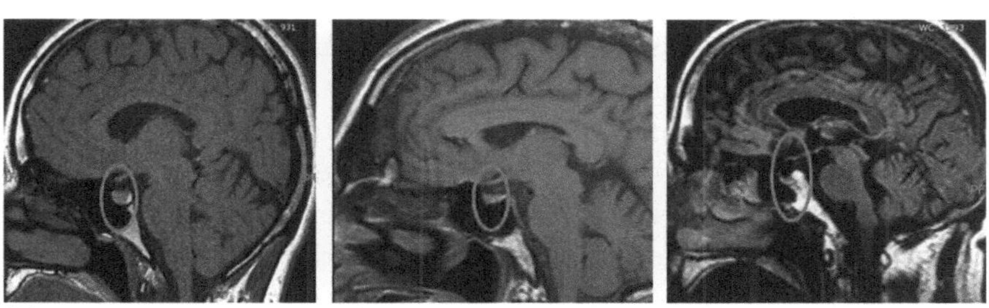

(A) (B) (C)

Figure 1 Three-dimensional MRI of a healthy control (A), 48 year normal control pituitary volume: 717 mm^3; GH-sufficient retired boxer (B), 46 year retired boxer without GH deficiency pituitary volume: 631 mm^3; and GH-deficient retired boxer (C), 47 year retired boxer with GH deficiency pituitary volume: 267 mm^3. GH-deficient retired boxer has significantly smaller pituitary volume than healthy control and GH-sufficient retired boxer.

examination and QoL-AGHDA questionnaires were administered for neuropsychologic tests and P300 ERPs were evaluated for cognitive performance. P300 amplitude, which is used as a measure of cognitive performance and provides a sensitive electrophysiologic index of the attentional and working memory demands of a task, was significantly lower in GH-deficient athletes when compared with GH-sufficient athletes and healthy controls.[53] Assessment of pituitary function in retired boxers and competing boxers who have a history of concussion is recommended.

5. COMBAT-RELATED TRAUMATIC BRAIN INJURY AND PITUITARY DYSFUNCTION

TBI has been commonly reported among soldiers participating in wars and it was demonstrated that it occurs in as much as 23% of combat soldiers.[54] Blast injury is the predominant type of TBI in combat soldiers.[54] Recently, war accidents have been reported as a new cause of neuroendocrine abnormalities. Hypopituitarism was reported as 25 and 42% among victims of combat-related TBI.[6,10] Wilkinson *et al.* conducted a cross-sectional study on 26 male combat veterans with blast-related mTBI whose last blast exposure was at least 1 year prior to the study. 42% of the study population with blast concussions were found to have abnormal hormone levels in one or more pituitary axes. The major limitation of that study is related to the diagnosis of hypopituitarism. Only basal hormone levels were measured and the participants did not undergo dynamic pituitary function tests.[6] In a recent study, Baxter *et al.* investigated the prevalence and consequences of pituitary dysfunction in 19 male soldiers with moderate to severe blast TBI. The soldiers underwent endocrine assessment including basal hormone measurements and dynamic pituitary function tests, namely the GST, GHRH-arginine test, ITT, and metyrapone test, between 2 and 48 months after injury. Hypopituitarism, GHD, and ACTH deficiency were diagnosed in 32, 16, and 10.5% of the soldiers, respectively. In that study, GH or ACTH deficiencies diagnosed by one dynamic test were confirmed with other tests.[55] The symptoms seen in posttraumatic stress disorder (PTSD) are similar to those seen in mTBI. Both may share the same symptoms and they co-occur. Of the patients with PTSD, 25% were reported to suffer from brain injury with or without loss of consciousness of short duration, less than 10 min.[56] In the assessment of PTSD patients in terms of neuroendocrine changes and symptom reports, mTBI should be considered because patients may have opposing biologic features which should also be taken into account in the planning of treatment for PTSD.[56] It is too early to recommend a screening strategy for pituitary functions in combat veterans because of the very limited data in the literature. In light of the current data, it could be recommended that war accident victims should be screened if they experience symptoms of hypopituitarism and were exposed to moderate to severe head injury including blast TBI.

6. CLINICAL MANIFESTATIONS OF TBI-INDUCED HYPOPITUITARISM

Patients with hypopituitarism may seek medical advice because of manifestations due to deficient hormone(s) or underlying causes, for example, visual defects due to pituitary adenoma compressing the optic chiasm. TBI as a cause of hypopituitarism cannot be revealed unless a careful medical history has been taken because the clinical presentation of hypopituitarism itself due to TBI does not differ from that seen in other causes. The recent guidelines released by the Endocrine Society suggested testing pituitary function in patients with TBI and SAH.[57] Patients who developed hypopituitarism after TBI present with manifestations due to the number of deficient hormones, severity of hormonal deficiency, and the duration of hypopituitarism without diagnosis and treatment. Independent of underlying cause, the clinical spectrum of hypopituitarism is very large and varies from some nonspecific complaints including fatigue, anorexia, arthralgia, and headache to life-threatening conditions which necessitate emergency admission such as adrenal crisis, hypothyroidism, and severe hypoglycemia.[58,59] Because of the mild and nonspecific manifestations recorded in the majority of patients with hypopituitarism due to different causes including TBI, most patients remain undiagnosed and thereby untreated. Since isolated GHD is the most common hormonal abnormality in patients with TBI, including sports-related head injury and the other causes, it would be expected that most patients would have clinical and biochemical findings compatible with GHD.[25,40,42] Hypothyroidism is associated with both disturbed neuropsychologic well being and neurocognitive functioning.[60] Hypogonadism is also associated with cognitive dysfunction.[61] Impaired cognitive function has been demonstrated in patients with both childhood-onset GHD and adult-onset GHD.[62,63] In a recent study conducted in athletes exposed to sports-related head injury, GHD after TBI was found to be associated with cognitive dysfunction, as revealed by decreased P300 amplitude, which is related to the updation of working memory content and attention.[53] Leon-Carrion *et al.* compared patients with isolated GHD after TBI with patients with TBI who had normal pituitary functions. They reported deficits in simple attention, increased reaction time, and greater emotional disruption in patients with isolated GHD due to TBI but not in control subjects.[64] It seems that the cognitive dysfunction seen following TBI is not only the result of brain injury itself but also of hypopituitarism. GHD, in particular, may result in cognitive abnormalities. GHD was shown to have adverse effects on executive abilities and mood in male veterans with mTBI.[10] Patients with GHD after TBI had reduced QoL in the domains of general health, physical health, and emotional health when compared to subjects with normal pituitary function.[65] Health-related quality of life (HRQL) and QoL-AGHDA in TBI survivors are worsened, independent of pituitary functions, when compared with healthy controls. In patients with TBI-induced hypopituitarism, they are significantly worsened when compared with TBI patients with normal pituitary function.[66] In the study by Popovic *et al.*, peak GH levels were significantly correlated with short- and long-term memory deficits in the patients with GHD due to TBI and there was also a correlation between lower IGF-I levels and visual memory.[67] Severe GHD may lead to lower IGF-I levels. It has been recently reported that decreased IGF-I levels seem to indicate an increased risk of developing Alzheimer's disease.[68] In an experimental study, serum IGF-I levels were found to be decreased in both the early and late periods of TBI and low levels of serum IGF-I were correlated with hippocampal neuron loss and spatial memory deficits.[69] GHD, which is the most common hormone deficiency in TBI victims, accounts for most of the clinical manifestations and biochemical abnormalities. One of the most important outcomes of childhood-onset GHD after TBI is related to height. Casanueva *et al.* compared patients with GHD after TBI with the patients with GHD due to nonfunctional pituitary adenoma (NFPA) in a Pfizer International Metabolic Database (KIMS) based observational study and they found that TBI-induced GH-deficient patients were significantly shorter than patients with GHD due to NFPA. The patients did not receive GH treatment before being included in KIMS.[70] Supporting data regarding height, the patients with childhood-onset GHD due to TBI were found to be significantly shorter than adult-onset GH-deficient patients. The patients were included in German KIMS database. The IGF-I SDS levels at baseline were also significantly lower in childhood-onset TBI patients than in adult-onset TBI patients.[71] In another recently reported KIMS based study, GH-deficient patients with TBI were found to be significantly younger and shorter than NFPA patients supporting the data reported previously.[72] All these data indicate that, at least in some of the subjects, short stature in adulthood might have resulted from TBI during childhood. Therefore, in the investigation of short stature in adults, any history of TBI should be investigated.

GHD in adults is associated with abnormal body composition, increased cardiovascular risk factors, abnormal lipid profile, altered bone metabolism, sleep and coagulation abnormalities.[73,74] Research on whether the metabolic and cardiovascular abnormalities seen in GHD due to other causes than TBI are similar to those seen in GHD due to TBI is insufficient. In a cross-sectional retrospective study, Prodam *et al.* reported that patients with TBI-induced hypopituitarism had a worse metabolic profile including insulin resistance, altered glucose levels, and hypertriglyceridemia than patients with TBI who had normal pituitary functions. In that study, GHD was the most common anterior hormone deficiency.[75] Tanriverdi *et al.* investigated anthropometric and body composition variables in 30 amateur boxers and in 15 healthy controls. They found that waist circumference, fat ratio, fat mass, abdominal fat ratio, abdominal fat mass, and serum leptin level were significantly high, and IGF-I level was significantly low in retired boxers. These findings indicate that retired boxers are at risk for cardiovascular complications.[76] Klose *et al.* reported that patients with hypopituitarism after TBI have higher BMI and abnormal lipid profile characterized by increased LDL-cholesterol and total cholesterol when compared to patients who have normal pituitary function.[66] Aerobic capacity in patients with TBI who have normal GH status is lower than normal aerobic capacity. When compared to GH-sufficient TBI survivors, aerobic capacity is worse in GH-deficient/insufficient survivors.[77]

7. PATHOPHYSIOLOGY OF TBI-INDUCED HYPOPITUITARISM

The mechanisms underlying the anterior pituitary dysfunction seen after TBI are still not completely understood. However, direct mechanical trauma to the hypothalamus, pituitary stalk or the pituitary gland, vascular/hypoxic insult to the hypothalamus or pituitary gland, inflammatory changes, compression from hemorrhage, edema or increased intracranial pressure, genetic predisposition, and autoimmunity may play a role in the development of neuroendocrine abnormalities.[78–82] Autopsy studies have given much valuable new insight into the pathophysiology of TBI-induced hypopituitarism. At least half of the autopsy studies preceded currently available modern methods used for the assessment of pituitary functions. Hypothalamo-pituitary damage was demonstrated in 26–86 of patients who died after TBI. The hypothalamus-pituitary region including posterior pituitary may be histologically/anatomically affected after TBI and hemorrhage: necrosis and fibrosis are the most commonly seen lesions. More than 70% of hemorrhagic infarctions were reported in the pituitary and 40% of hypothalamic microhemorrhages were reported in the hypothalamus in autopsy studies.[78,79,83–86] Harper *etal.* examined the pathologic findings of pituitaries obtained from a consecutive series of 100 patients who died as a result of nonmissile TBI and the most common cause of TBI was road traffic accidents. The most common abnormality in the pituitary gland was infarction in the anterior lobe; infarcts were observed in 38% of the pituitaries.[80] Massive infarction of the anterior lobe is the result of shock following trauma combined with anoxia.[87] Interestingly, the patients with massive infarction had suffered a cardiorespiratory arrest and had required artificial ventilation for more than 12 h before their death.[80,87] Our recent data support this interesting autopsy finding. The frequencies of GH, gonadotropin, and TSH deficiencies were found as 27.2, 9.0, and 2.2%, respectively in patients who had ventricular arrhythmias 12–36 months previously.[88]

It seems that the development of anterior pituitary dysfunction after TBI is related to the anatomy of the hypothalamus-pituitary region. Very peculiar anatomic and special vascular features make the hypothalamus and pituitary more vulnerable to direct mechanical trauma and ischemic/hypoxic changes. About 70–90% of the blood supply to the anterior lobe of the pituitary gland is supplied by branches from the long hypophyseal portal system, which takes blood supply from the superior hypophyseal artery. These branches originate from above the diaphragma and pass through the diaphragmatic sella. The short hypophyseal vessels that come from below the diaphragma supply blood to the central portion of the gland. The somatotrophs and gonadotrophs are found more commonly in the lateral wings of the pituitary gland.[89] Since the somatotroph cells perfused by the long hypophyseal portal system may be damaged after head trauma more easily, it is not surprising that GH is the most common deficient hormone after TBI.

APO E is a key protein which is responsible for enhancing lipid transport and metabolism within the nervous system. It is one of the most abundant proteins in the hypothalamus-pituitary region and may also play a role in neuronal repair mechanisms. There is a potential association between APO E polymorphism and head injury outcome.[90] The relation of the APO E ε4 allele with chronic traumatic encephalopathy was also demonstrated.[91] We have demonstrated for the first time that APO E polymorphism is associated with the development of TBI-induced pituitary dysfunction. Since less pituitary dysfunction occurred in TBI victims who had the APO E3/E3 genotype, genetic predisposition may be an important factor in the development of hypopituitarism.[82] As mentioned previously, pituitary dysfunction may improve over time in a considerable number of patients but, rarely, it may also worsen over the 5-year period or new-onset isolated hypopituitarism may appear over the years.[40] A possible role of autoimmunity was demonstrated previously in animals exposed to experimental TBI.[92] Then, the possible role of autoimmunity in the pituitary dysfunction after TBI was studied in patients with TBI and athletes participating in combative sports. APAs and AHAs were measured in patients and healthy controls. These studies revealed that autoimmunity may have a role in the development and/or worsening of hypopituitarism. More commonly hypopituitarism was found in patients with TBI and athletes who had higher APAs and AHAs.[81] Although the presence of both APAs and AHAs is associated with hypopituitarism after noncombative TBI, AHAs are associated with pituitary dysfunction in boxers.[51] Tanycytes which are special ependymal cells are responsible for bridging the cerebrospinal fluid with the portal capillaries. In an experimental TBI model, it was shown that the barrier properties of tanycytes were compromised by the disruption of tight junctions, which may contribute to the development of pituitary dysfunction.[93] The recovery of somatotrope cells destroyed by diphteria toxin, another type of injury, from stem/progenitor cells gives a very promising therapeutic approach in the future.[94]

8. SCREENING STRATEGY

Since the importance of hypopituitarism after TBI has only been recognized recently, the medical community was totally unaware that TBI could have been a significant cause of hypopituitarism and questioning of TBI was not included in medical history targeted to clarify the underlying cause of hypopituitarism. Unawareness of hypopituitarism after TBI is one of the most likely reasons for ignoring the problem and leaving the patient undiagnosed and thereby untreated.

The pituitary function is disturbed in TBI in acute phase of the event.[95] Most of the hormonal changes seen in acute phase are physiologic responses to critical illness and their routine screening is not recommended. Since cortisol deficiency is life-threatening, ACTH deficiency should be checked routinely, especially in patients with clinical context compatible with adrenal failure. In the case of hypocortisolism, glucocorticoid replacement is required. However, hypothalamus-pituitary axis needs to be checked after the patient was discharged from emergency department and clinical and neurological condition were stabilized.

According to the current data, the most appropriate time for endocrine assessment is 1 year after TBI and the best TBI victims for screening are those with moderate or severe TBI.[96] The patients with severe disability or vegetative state are not adequate candidates for screening of pituitary function, because they do not benefit from replacement therapy. Testing of all mTBI patients is not cost effective. Current data suggest that patients with mTBI who require hospitalization for at least 24 h, who have radiological abnormalities on initial CT, who have longer duration of intubation and hospitalisation in intensive care unit, and who have manifestations of hypopitutarism should be screened at any time after TBI.[95] As mentioned previously, most of the anterior pituitary deficiencies improve over time, but still there are substantial percentages of hypopituitarism in the 5th year.[95] Therefore, the authors recommend that the patients, independent of the severity of trauma, should be followed-up annually for at least 5 years, as carrying out only medical history and physical examination as the minimal required procedure. Only patients with clinical picture, that is compatible with hypopituitarism, are advised for screening of pituitary function.

9. TREATMENT OF HYPOPITUITARISM IN PATIENTS WITH TBI

The long-term effects of replacement therapy in patients with TBI-induced hypopituitarism are not available yet. The current data depend on case reports, small series including a small number of patients and a few retrospective studies. The treatment of TBI-induced hypopituitarism depends on the appropriate replacement of the deficient hormone(s) and is not different from the treatment of hypopituitarism due to other causes. Reversibility of pituitary dysfunction in TBI victims should be kept in mind before treatment is started, which may be life-long, such as in hypothyroidism or hypoadrenalism, or long term, such as in GHD. Besides cost-effectiveness, replacement therapy is not without side effects and careful monitoring is necessary. Since the most common deficient hormone is GH and isolated GHD is not uncommon, most of the studies published so far in the literature are related to the effects of GH replacement. rhGH replacement therapy improves muscle force production, body composition, and aerobic capacity.[97] The beneficial effects of GHRT on QoL have been shown in a retrospective German KIMS study including 84 patients with TBI and 84 patients with nonfunctioning pituitary adenoma and it was shown that QoL was significantly improved in both groups after GH replacement therapy.[71] In addition to the previously mentioned nation-based KIMS study, recently Gardner *et al.* retrospectively assessed the clinical characteristics of hypopituitarism due to TBI and the QoL measured by QoL-AGHDA scores and the effects of long-term GH replacement therapy in a large number of TBI patients involved in the KIMS database. Patients with NFPA were used as the control group. When compared to NFPA, hypopituitarism in TBI victims was biochemically less severe than hypopituitarism due to NFPA but QoL was worse in TBI victims and GH replacement therapy not only significantly improved QoL but sustained the relevant improvement in the long-term period in patients with TBI. Although retrospective, this is the most comprehensive study providing an evidence based for the treatment of GHD in TBI-induced hypopituitarism.[72] Obviously, more studies are required to understand the effectiveness of hormone replacement therapies in patients with hypopituitarism due to TBI.

10. CONCLUSIONS

TBI is one of the common causes of anterior hypopituitarism and GHD, in particular. The real prevalence of TBI-induced hypopituitarism is not well known and markedly differs between the studies due to unawareness of the problem, the subtle clinical manifestations of mild hypopituitarism, the type and severity of trauma, number of trauma, study population, the design of the study, the diagnostic tests, and criteria of hormone deficiencies. Not uncommonly, diagnosis of TBI-induced hypopituitarism is usually delayed and the condition remains untreated. The underlying mechanisms responsible for the development of pituitary dysfunction after TBI are not completely clarified yet. Direct mechanical trauma and vascular/hypoxic insult to the hypothalamus, pituitary stalk or pituitary gland, compression from hemorrhage, edema or increased intracranial pressure are some factors proposed to explain the pathophysiology of TBI-induced pituitary dysfunction. Genetic and immunologic studies have given new insight into the pathophysiology of TBI-induced pituitary dysfunction. The APO E3/E3 genotype decreases the risk of TBI-induced hypopituitarism. APO E may have some neuroprotective and neurodegenerative properties in the hypothalamus-pituitary region and it may have a significant impact in terms of the diagnosis and treatment of pituitary dysfunction after TBI. On the other hand, autoimmunity may have a role in the development and/or worsening of hypopituitarism in patients with TBI. Both APAs and AHAs are associated with hypopituitarism in patients with TBI and persist even 5 years after TBI.

Sports-related injury may cause pituitary dysfunction. The medical community, not only endocrine community, should be alert to pituitary dysfunction in athletes participating in combative sports such as boxing, and in martial arts such as kickboxing. Boxing and kickboxing are characterized by chronic repetitive head trauma. Boxing duration, number of bouts, and the presence of AHAs are associated with the development of pituitary dysfunction in athletes.

The current data regarding both war accidents and sports-related injury-induced pituitary dysfunctions are not sufficient to recommend a screening policy. The screening strategy obtained in a noncombative population may be applied to a military population and athletes until more data become available.

In the treatment of hypopituitarism due to TBI, the deficient hormones should be appropriately replaced after hypopituitarism has been diagnosed as permanent. GH replacement therapy in GH-deficient patients who had combat-related or noncombat related-TBI not only improves QoL but also improves metabolic abnormalities including glucose intolerance and abnormal lipid profile. Experimental data regarding stem-cell therapy in hypopituitarism due to injured pituitary are hopeful.

REFERENCES

1. Corrigan JD, Selassie AW, Orman JA. The epidemiology of traumatic brain injury. *J Head Trauma Rehabil.* 2010;25:72–80.
2. Tagliaferri F, Compagnone C, Korsic M, Servadei F, Kraus J. A systematic review of brain injury epidemiology in Europe. *Acta Neurochir.* 2006;148:255–268; discussion 268.
3. Hillier SL, Hiller JE, Metzer J. Epidemiology of traumatic brain injury in South Australia. *Brain Inj.* 1997;11:649–659.
4. McKinlay A, Grace RC, Horwood LJ, Fergusson DM, Ridder EM, MacFarlane MR. Prevalence of traumatic brain injury among children, adolescents and young adults: prospective evidence from a birth cohort. *Brain Inj.* 2008;22:175–181.

5. Consensus Conference.Rehabilitation of persons with traumatic brain injury. NIH Consensus Development Panel on rehabilitation of persons with traumatic brain injury. *JAMA.* 1999;282:974–983.
6. Wilkinson CW, Pagulayan KF, Petrie EC, et al. High prevalence of chronic pituitary and target-organ hormone abnormalities after blast-related mild traumatic brain injury. *Front Neurol.* 2012;3:11.
7. Cyran E. Hypophysenschadigung durch schadelbasisfraktur. *Dtsch Med Wschr.* 1918;44: 1261.
8. Benvenga S, Campenni A, Ruggeri RM, Trimarchi F. Clinical review 113: hypopituitarism secondary to head trauma. *J Clin Endocrinol Metab.* 2000;85:1353–1361.
9. Kelestimur F. Chronic trauma in sports as a cause of hypopituitarism. *Pituitary.* 2005;8: 259–262.
10. Ioachimescu AG, Hampstead BM, Moore A, Burges E, Phillips LS. Growth hormone deficiency after combat-related traumatic brain injury. *Pituitary.* 2015;18:535–541.
11. Tanriverdi F, Ulutabanca H, Unluhizarci K, Selcuklu A, Casanueva FF, Kelestimur F. Pituitary functions in the acute phase of traumatic brain injury: are they related to severity of the injury or mortality? *Brain Inj.* 2007;21:433–439.
12. Tanriverdi F, Senyurek H, Unluhizarci K, Selcuklu A, Casanueva FF, Kelestimur F. High risk of hypopituitarism after traumatic brain injury: a prospective investigation of anterior pituitary function in the acute phase and 12 months after trauma. *J Clin Endocrinol Metab.* 2006;91:2105–2111.
13. Klose M, Juul A, Struck J, Morgenthaler NG, Kosteljanetz M, Feldt-Rasmussen U. Acute and long-term pituitary insufficiency in traumatic brain injury: a prospective single-centre study. *Clin Endocrinol.* 2007;67:598–606.
14. Prasanna KL, Mittal RS, Gandhi A. Neuroendocrine dysfunction in acute phase of moderate-to-severe traumatic brain injury: a prospective study. *Brain Inj.* 2015;29:336–342.
15. Hannon MJ, Crowley RK, Behan LA, et al. Acute glucocorticoid deficiency and diabetes insipidus are common after acute traumatic brain injury and predict mortality. *J Clin Endocrinol Metab.* 2013;98:3229–3237.
16. Cooper MS, Stewart PM. Adrenal insufficiency in critical illness. *J Intensive Care Med.* 2007;22:348–362.
17. Agha A, Phillips J, O'Kelly P, Tormey W, Thompson CJ. The natural history of post-traumatic hypopituitarism: implications for assessment and treatment. *Am J Med.* 2005;118:1416:e1–e7.
18. Kelly DF, Gonzalo IT, Cohan P, Berman N, Swerdloff R, Wang C. Hypopituitarism following traumatic brain injury and aneurysmal subarachnoid hemorrhage: a preliminary report. *J Neurosurg.* 2000;93:743–752.
19. Lieberman SA, Oberoi AL, Gilkison CR, Masel BE, Urban RJ. Prevalence of neuroendocrine dysfunction in patients recovering from traumatic brain injury. *J Clin Endocrinol Metab.* 2001;86:2752–2756.
20. van der Eerden AW, Twickler MT, Sweep FC, et al. Should anterior pituitary function be tested during follow-up of all patients presenting at the emergency department because of traumatic brain injury? *Eur J Endocrinol.* 2010;162:19–28.
21. Bavisetty S, Bavisetty S, McArthur DL, et al. Chronic hypopituitarism after traumatic brain injury: risk assessment and relationship to outcome. *Neurosurgery.* 2008;62: 1080–1093; discussion 1093–1084.
22. Bondanelli M, De Marinis L, Ambrosio MR, et al. Occurrence of pituitary dysfunction following traumatic brain injury. *J Neurotrauma.* 2004;21:685–696.
23. Kokshoorn NE, Smit JW, Nieuwlaat WA, et al. Low prevalence of hypopituitarism after traumatic brain injury: a multicenter study. *Eur J Endocrinol.* 2011;165:225–231.
24. Agha A, Rogers B, Sherlock M, et al. Anterior pituitary dysfunction in survivors of traumatic brain injury. *J Clin Endocrinol Metab.* 2004;89:4929–4936.

25. Aimaretti G, Ambrosio MR, Di Somma C, et al. Residual pituitary function after brain injury-induced hypopituitarism: a prospective 12-month study. *J Clin Endocrinol Metab.* 2005;90:6085–6092.
26. Herrmann BL, Rehder J, Kahlke S, et al. Hypopituitarism following severe traumatic brain injury. *Exp Clin Endocrinol Diabetes.* 2006;114:316–321.
27. Leal-Cerro A, Flores JM, Rincon M, et al. Prevalence of hypopituitarism and growth hormone deficiency in adults long-term after severe traumatic brain injury. *Clin Endocrinol.* 2005;62:525–532.
28. Schneider HJ, Schneider M, Saller B, et al. Prevalence of anterior pituitary insufficiency 3 and 12 months after traumatic brain injury. *Eur J Endocrinol.* 2006;154:259–265.
29. Schneider M, Schneider HJ, Yassouridis A, Saller B, von Rosen F, Stalla GK. Predictors of anterior pituitary insufficiency after traumatic brain injury. *Clin Endocrinol.* 2008;68: 206–212.
30. Jeong JH, Kim YZ, Cho YW, Kim JS. Negative effect of hypopituitarism following brain trauma in patients with diffuse axonal injury. *J Neurosurg.* 2010;113:532–538.
31. Klose M, Juul A, Poulsgaard L, Kosteljanetz M, Brennum J, Feldt-Rasmussen U. Prevalence and predictive factors of post-traumatic hypopituitarism. *Clin Endocrinol.* 2007;67:193–201.
32. Wachter D, Gündling K, Oertel MF, Stracke H, Böker D-K. Pituitary insufficieny after traumatic brain injury. *J Clin Neurosci.* 2009;16:201–208.
33. Berg C, Oeffner A, Schumm-Draeger P-M, Badorrek F, Brabant G, Gerbert B, Bornstein S, Zimmermann A, Weber M, Broecker-Preuss M, Mann K, Herrmann BL. Prevalence of anterior pituitary dysfunction in patients following traumatic brain injury in a German multi-centre screening programme. *Exp Clin Endocrinol Diabetes.* 2010;118:139–144.
34. Ulfarsson T, Arnar Gudnason G, Rosen T, et al. Pituitary function and functional outcome in adults after severe traumatic brain injury: the long-term perspective. *J Neurotrauma.* 2013;30:271–280.
35. Nemes O, Kovacs N, Czeiter E, et al. Predictors of post-traumatic pituitary failure during long-term follow-up. *Hormones.* 2014;14:383–391.
36. Ulutabanca H, Hatipoglu N, Tanriverdi F, et al. Prospective investigation of anterior pituitary function in the acute phase and 12 months after pediatric traumatic brain injury. *Childs Nerv Syst.* 2014;30:1021–1028.
37. Schneider HJ, Kreitschmann-Andermahr I, Ghigo E, Stalla GK, Agha A. Hypothalamopituitary dysfunction following traumatic brain injury and aneurysmal subarachnoid hemorrhage: a systematic review. *JAMA.* 2007;298:1429–1438.
38. Kokshoorn NE, Wassenaar MJ, Biermasz NR, et al. Hypopituitarism following traumatic brain injury: prevalence is affected by the use of different dynamic tests and different normal values. *Eur J Endocrinol.* 2010;162:11–18.
39. Klose M, Stochholm K, Janukonyte J, et al. Prevalence of posttraumatic growth hormone deficiency is highly dependent on the diagnostic set-up: results from The Danish National Study on Posttraumatic Hypopituitarism. *J Clin Endocrinol Metab.* 2014;99: 101–110.
40. Tanriverdi F, De Bellis A, Ulutabanca H, et al. A five year prospective investigation of anterior pituitary function after traumatic brain injury: is hypopituitarism long-term after head trauma associated with autoimmunity? *J Neurotrauma.* 2013;30:1426–1433.
41. Tanriverdi F, Ulutabanca H, Unluhizarci K, Selcuklu A, Casanueva FF, Kelestimur F. Three years prospective investigation of anterior pituitary function after traumatic brain injury: a pilot study. *Clin Endocrinol.* 2008;68:573–579.
42. Kelestimur F, Tanriverdi F, Atmaca H, Unluhizarci K, Selcuklu A, Casanueva FF. Boxing as a sport activity associated with isolated GH deficiency. *J Endocrinol Invest.* 2004;27: RC28–RC32.

43. Kelly DF, Chaloner C, Evans D, et al. Prevalence of pituitary hormone dysfunction, metabolic syndrome, and impaired quality of life in retired professional football players: a prospective study. *J Neurotrauma.* 2014;31:1161–1171.
44. Tanriverdi F, Unluhizarci K, Coksevim B, Selcuklu A, Casanueva FF, Kelestimur F. Kickboxing sport as a new cause of traumatic brain injury-mediated hypopituitarism. *Clin Endocrinol.* 2007;66:360–366.
45. Tanriverdi F, Unluhizarci K, Kocyigit I, et al. Brief communication: pituitary volume and function in competing and retired male boxers. *Ann Intern Med.* 2008;148:827–831.
46. Ohhashi G, Tani S, Murakami S, Kamio M, Abe T, Ohtuki J. Problems in health management of professional boxers in Japan. *Br J Sports Med.* 2002;36:346–352; discussion 353.
47. Zhang L, Ravdin LD, Relkin N, et al. Increased diffusion in the brain of professional boxers: a preclinical sign of traumatic brain injury? *Am J Neuroradiol.* 2003;24:52–57.
48. Jordan BD. Chronic traumatic brain injury associated with boxing. *Semin Neurol.* 2000;20:179–185.
49. Gartland S, Malik MH, Lovell ME. Injury and injury rates in Muay Thai kick boxing. *Br J Sports Med.* 2001;35:308–313.
50. Tanriverdi F, Unluhizarci K, Selcuklu A, Casanueva FF, Kelestimur F. Transient hypogonadotropic hypogonadism in an amateur kickboxer after head trauma. *J Endocrinol Invest.* 2007;30:150–152.
51. Tanriverdi F, De Bellis A, Battaglia M, et al. Investigation of antihypothalamus and antipituitary antibodies in amateur boxers: is chronic repetitive head trauma-induced pituitary dysfunction associated with autoimmunity? *Eur J Endocrinol.* 2010;162:861–867.
52. Dupuis F, Johnston KM, Lavoie M, Lepore F, Lassonde M. Concussions in athletes produce brain dysfunction as revealed by event-related potentials. *Neuroreport.* 2000;11:4087–4092.
53. Tanriverdi F, Suer C, Yapislar H, et al. Growth hormone deficiency due to sports-related head trauma is associated with impaired cognitive performance in amateur boxers and kickboxers as revealed by P300 auditory event-related potentials. *Clin Endocrinol.* 2013;78:730–737.
54. Guerrero AF, Alfonso A. Traumatic brain injury-related hypopituitarism: a review and recommendations for screening combat veterans. *Mil Med.* 2010;175:574–580.
55. Baxter D, Sharp DJ, Feeney C, et al. Pituitary dysfunction after blast traumatic brain injury: the UK BIOSAP study. *Ann Neurol.* 2013;74:527–536.
56. Flory JD, Henn-Haase C, Bierer LM, et al. Glucocorticoid functioning in male combat veterans with posttraumatic stress disorder and mild traumatic brain injury. *Biol Psychiatry.* 2015;78:e5–e6.
57. Molitch ME, Clemmons DR, Malozowski S, Merriam GR, Vance ML. Endocrine Society.Evaluation and treatment of adult growth hormone deficiency: an Endocrine Society clinical practice guideline. *J Clin Endocrinol Metab.* 2011;96:1587–1609.
58. Diri H, Tanriverdi F, Karaca Z, et al. Extensive investigation of 114 patients with Sheehan's syndrome: a continuing disorder. *Eur J Endocrinol.* 2014;171:311–318.
59. Guven M, Bayram F, Guven K, Kelestimur F. Evaluation of patients admitted with hypoglycaemia to a teaching hospital in Central Anatolia. *Postgrad Med J.* 2000;76:150–152.
60. Wekking EM, Appelhof BC, Fliers E, et al. Cognitive functioning and well-being in euthyroid patients on thyroxine replacement therapy for primary hypothyroidism. *Eur J Endocrinol.* 2005;153:747–753.
61. Cherrier MM, Craft S, Matsumoto AH. Cognitive changes associated with supplementation of testosterone or dihydrotestosterone in mildly hypogonadal men: a preliminary report. *J Androl.* 2003;24:568–576.
62. Deijen JB, de Boer H, Blok GJ, van der Veen EA. Cognitive impairments and mood disturbances in growth hormone deficient men. *Psychoneuroendocrinology.* 1996;21:313–322.

63. Golgeli A, Tanriverdi F, Suer C, et al. Utility of P300 auditory event related potential latency in detecting cognitive dysfunction in growth hormone (GH) deficient patients with Sheehan's syndrome and effects of GH replacement therapy. *Eur J Endocrinol.* 2004;150:153–159.
64. Leon-Carrion J, Leal-Cerro A, Cabezas FM, et al. Cognitive deterioration due to GH deficiency in patients with traumatic brain injury: a preliminary report. *Brain Inj.* 2007;21:871–875.
65. Kelly DF, McArthur DL, Levin H, et al. Neurobehavioral and quality of life changes associated with growth hormone insufficiency after complicated mild, moderate, or severe traumatic brain injury. *J Neurotrauma.* 2006;23:928–942.
66. Klose M, Watt T, Brennum J, Feldt-Rasmussen U. Posttraumatic hypopituitarism is associated with an unfavorable body composition and lipid profile, and decreased quality of life 12 months after injury. *J Clin Endocrinol Metab.* 2007;92:3861–3868.
67. Popovic V, Pekic S, Pavlovic D, et al. Hypopituitarism as a consequence of traumatic brain injury (TBI) and its possible relation with cognitive disabilities and mental distress. *J Endocrinol Invest.* 2004;27:1048–1054.
68. Westwood AJ, Beiser A, Decarli C, et al. Insulin-like growth factor-1 and risk of Alzheimer dementia and brain atrophy. *Neurology.* 2014;82:1613–1619.
69. Ozdemir D, Baykara B, Aksu I, et al. Relationship between circulating IGF-1 levels and traumatic brain injury-induced hippocampal damage and cognitive dysfunction in immature rats. *Neurosci Lett.* 2012;507:84–89.
70. Casanueva FF, Leal A, Koltowska-Haggstrom M, Jonsson P, Goth MI. Traumatic brain injury as a relevant cause of growth hormone deficiency in adults: a KIMS-based study. *Arch Phys Med Rehabil.* 2005;86:463–468.
71. Kreitschmann-Andermahr I, Poll EM, Reineke A, et al. Growth hormone deficient patients after traumatic brain injury—baseline characteristics and benefits after growth hormone replacement—an analysis of the German KIMS database. *Growth Horm IGF Res.* 2008;18:472–478.
72. Gardner CJ, Mattsson AF, Daousi C, Karbonits M, Koltowska-Haggstrom M, Cuthbertson D. Growth hormone deficiency after traumatic brain injury: improvement in quality of life with GH therapy-analysis of the KIMS database. *Eur J Endocrinol.* 2015;172:371–381.
73. Gola M, Bonadonna S, Doga M, Giustina A. Clinical review: growth hormone and cardiovascular risk factors. *J Clin Endocrinol Metab.* 2005;90:1864–1870.
74. Tanriverdi F, Karaca Z, Unluhizarci K, Kelestimur F. Unusual effects of GH deficiency in adults: a review about the effects of GH on skin, sleep, and coagulation. *Endocrine.* 2014;47:679–689.
75. Prodam F, Gasco V, Caputo M, et al. Metabolic alterations in patients who develop traumatic brain injury (TBI)-induced hypopituitarism. *Growth Horm IGF Res.* 2013;23: 109–113.
76. Tanriverdi F, Kocyigit I, Unluhizarci K, Casanueva FF, Kelestimur F. Body composition, serum IGF-I and leptin level changes in amateur boxers: retired boxers have risk factors for cardiovascular disorders. *Obes Metab.* 2008;4:118–123.
77. Mossberg KA, Masel BE, Gilkison CR, Urban RJ. Aerobic capacity and growth hormone deficiency after traumatic brain injury. *J Clin Endocrinol Metab.* 2008;93: 2581–2587.
78. Ceballos R. Pituitary changes in head trauma (analysis of 102 consecutive cases of head injury). *Ala J Med Sci.* 1966;3:185–198.
79. Crompton MR. Hypothalamic lesions following closed head injury. *Brain.* 1971;94: 165–172.
80. Harper CG, Doyle D, Adams JH, Graham DI. Analysis of abnormalities in pituitary gland in non-missile head injury: study of 100 consecutive cases. *J Clin Pathol.* 1986;39:769–773.

81. Tanriverdi F, De Bellis A, Bizzarro A, et al. Antipituitary antibodies after traumatic brain injury: is head trauma-induced pituitary dysfunction associated with autoimmunity? *Eur J Endocrinol.* 2008;159:7–13.
82. Tanriverdi F, Taheri S, Ulutabanca H, et al. Apolipoprotein E3/E3 genotype decreases the risk of pituitary dysfunction after traumatic brain injury due to various causes: preliminary data. *J Neurotrauma.* 2008;25:1071–1077.
83. Daniel PM, Prichard MM, Treip CS. Traumatic infarction of the anterior lobe of the pituitary gland. *Lancet.* 1959;2:927–931.
84. Kornblum RN, Fisher RS. Pituitary lesions in craniocerebral injuries. *Arch Pathol.* 1969;88:242–248.
85. Pierucci G, Gherson G, Tavani M. Pituitary changes especially necrotic—following cranio-cerebral injuries. *Pathologica.* 1971;63:71–88.
86. Salehi F, Kovacs K, Scheithauer BW, Pfeifer EA, Cusimano M. Histologic study of the human pituitary gland in acute traumatic brain injury. *Brain Inj.* 2007;21:651–656.
87. Orthner H, Meyer E. Der posttraumatische diabetes insipidus. Befunde am neurosekretorischen system beim stumpfen schaldeltrauma, nebst bemerkungen zum posttraumatischen hirnödem. *Acta Neuroveg (Wien).* 1967;30:216–250.
88. Simsek Y, Kaya MG, Tanriverdi F, et al. Evaluation of long-term pituitary functions in patients with severe ventricular arrhythmia: a pilot study. *J Endocrinol Invest.* 2014;37: 1057–1064.
89. Gorczyca W, Hardy J. Arterial supply of the human anterior pituitary gland. *Neurosurgery.* 1987;20:369–378.
90. Nicoll JA, Roberts GW, Graham DI. Apolipoprotein E epsilon 4 allele is associated with deposition of amyloid beta-protein following head injury. *Nature Med.* 1995;1:135–137.
91. Jordan BD, Relkin NR, Ravdin LD, Jacobs AR, Bennett A, Gandy S. Apolipoprotein E epsilon4 associated with chronic traumatic brain injury in boxing. *JAMA.* 1997;278: 136–140.
92. Rudehill S, Muhallab S, Wennersten A, et al. Autoreactive antibodies against neurons and basal lamina found in serum following experimental brain contusion in rats. *Acta Neurochir.* 2006;148:199–205; discussion 205.
93. Osterstock G, El Yandouzi T, Romano N, et al. Sustained alterations of hypothalamic tanycytes during posttraumatic hypopituitarism in male mice. *Endocrinology.* 2014;155: 1887–1898.
94. Fu Q, Gremeaux L, Luque RM, et al. The adult pituitary shows stem/progenitor cell activation in response to injury and is capable of regeneration. *Endocrinology.* 2012;153: 3224–3235.
95. Tanriverdi F, Unluhizarci K, Kelestimur F. Pituitary function in subjects with mild traumatic brain injury: a review of literature and proposal of a screening strategy. *Pituitary.* 2010;13:146–153.
96. Fernandez-Rodriguez E, Bernabeu I, Castro AI, Kelestimur F, Casanueva FF. Hypopituitarism following traumatic brain injury determining factors for diagnosis. *Front Endocrinol.* 2011;2:25.
97. Bhagia V, Gilkison C, Fitts RH, et al. Effect of recombinant growth hormone replacement in a growth hormone deficient subject recovering from mild traumatic brain injury: a case report. *Brain Inj.* 2010;24:560–567.

CHAPTER NINE

Effects of Growth Hormone on Bone

Nicholas A. Tritos[*,†,1], **Anne Klibanski**[*,†]
[*] Neuroendocrine Unit, Massachusetts General Hospital, Boston, Massachusetts, USA
[†] Harvard Medical School, Boston, Massachusetts, USA
[1] Corresponding author: email address: ntritos@mgh.harvard.edu.

Contents

Abstract

Purpose

Describe the effects of growth hormone (GH) and insulin-like growth factor 1 (IGF-1) on the skeleton.

Findings

The GH and IGF-1 axis has pleiotropic effects on the skeleton throughout the lifespan by influencing bone formation and resorption. GH deficiency leads to decreased bone turnover, delayed statural growth in children, low bone mass, and increased fracture risk in adults. GH replacement improves adult stature in GH deficient children, increases bone mineral density (BMD) in adults, and helps to optimize peak bone acquisition in patients, during the transition from adolescence to adulthood, who have persistent GH deficiency. Observational studies suggest that GH replacement may mitigate the excessive fracture risk associated with GH deficiency.

Acromegaly, a state of GH and IGF-1 excess, is associated with increased bone turnover and decreased BMD in the lumbar spine observed in some studies, particularly in patients with hypogonadism. In addition, patients with acromegaly appear to be at an increased risk of morphometric-vertebral fractures, especially in the presence of active disease or concurrent hypogonadism.

Progress in Molecular Biology and Translational Science, Volume 138
ISSN 1877-1173
http://dx.doi.org/10.1016/bs.pmbts.2015.10.008

GH therapy also has beneficial effects on statural growth in several conditions characterized by GH insensitivity, including chronic renal failure, Turner syndrome, Prader–Willi syndrome, postnatal growth delay in patients with intrauterine growth retardation who do not demonstrate catchup growth, idiopathic short stature, short stature homeobox-containing (*SHOX*) gene mutations, and Noonan syndrome.

Summary
GH and IGF-1 have important roles in skeletal physiology, and GH has an important therapeutic role in both GH deficiency and insensitivity states.

1. INTRODUCTION

Our understanding of the effects of growth hormone (GH) on bone has vastly increased over the past few decades. Originally isolated as a substance promoting linear growth in young animals and humans during childhood and adolescence, GH was subsequently shown to exert positive effects on calcium balance in the mature skeleton of dogs in early studies by Harris and Heaney.[1] The availability of recombinant human GH, has significantly facilitated studies of its effects on the skeleton throughout the lifespan.[2]

This review summarizes the findings of preclinical and clinical studies on the effects of GH on bone, emphasizing recent developments in the field and highlighting directions for future study.

2. PRECLINICAL DATA

GH is secreted by pituitary somatotrophs in a pulsatile manner and acts on peripheral tissues, either directly or indirectly, through the stimulation of insulin-like growth factor 1 (IGF-1) synthesis and secretion (as described in the currently well-established "somatomedin hypothesis").[3,4] In addition to the direct actions of GH, both IGF-1 of endocrine origin, predominantly secreted by the liver into the systemic circulation, and locally synthesized IGF-1 acting in a paracrine manner, appear to be relevant to the GH actions in the skeleton (Fig. 1). Mice with genetic deficiencies of the GH receptor and IGF-1, have greater decreases in bone length compared to animals having either deficiency alone, suggesting that each molecule has an important role in bone biology.[5]

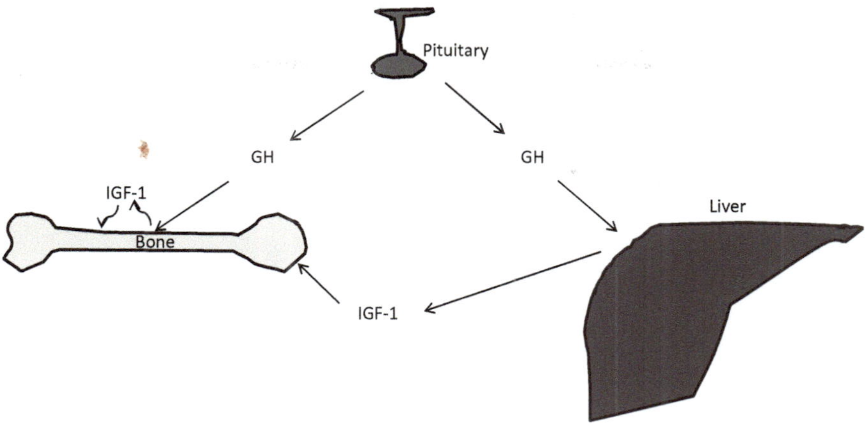

Figure 1 Growth hormone, insulin-like growth factor 1 (IGF-1), and the skeleton. Abbreviations: GH, growth hormone; and IGF-1, insulin-like growth factor 1.

The effects of GH and IGF-1 on epiphyseal plates have been extensively studied and appear to be species-specific. In rodent chondrocytes, IGF-1 is synthesized under GH stimulation, and acts in a paracrine manner to induce chondrocyte proliferation and endochondral ossification, leading to linear bone growth.[6] In contrast, in bovine epiphyseal cartilage, systemic (rather than paracrine) IGF-1 acts to stimulate growth, as very little IGF-1 is locally synthesized in response to GH.[7] In addition, basic fibroblast growth factor acts on epiphyseal plates to increase the number of IGF-1 receptors, thus enhancing tissue response to IGF-1 of endocrine origin.[7]

Based on the data obtained from animal models, systemic IGF-1 appears to predominantly influence cortical bone. Mice with genetic GH receptor deficiency, have decreased cortical thickness without apparent impairment of trabecular bone thickness and volume, a phenotype that is rescued by systemic IGF-1 administration.[8] In contrast, paracrine IGF-1 appears to have a predominant effect on trabecular bone. Mice with transgenic overexpression of IGF-1 in osteoblasts, have increased trabecular bone volume, whereas osteoblast-specific IGF-1 knockout mice have decreased trabecular volume and impaired trabecular structure and mineralization.[9,10]

GH directly stimulates osteoblastogenesis and bone formation.[11] In addition, osteoblasts express IGF-1 under parathyroid hormone (PTH) stimulation.[12,13] In turn, IGF-1 mediates the anabolic effects of intermittent PTH administration on bone by promoting bone formation.[14] Several other hormones, including thyroid hormone and estrogen also stimulate paracrine IGF-1 secretion by osteoblasts, whereas glucocorticoids inhibit it.[15–17]

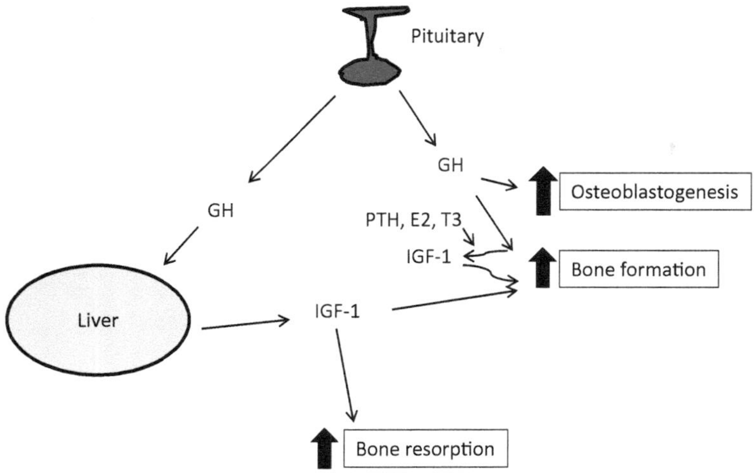

Figure 2 Effects of growth hormone, insulin-like growth factor 1 (IGF-1), and other hormones on skeletal tissue. Abbreviations: E2, estradiol; GH, growth hormone; IGF-1, insulin-like growth factor 1; PTH, parathyroid hormone; and T3, triiodothyronine.

Furthermore, several IGF binding proteins (IGFBPs) are locally expressed in bone and modulate IGF-1 action.[18,19] In addition to its effects on bone formation, IGF-1 induces the synthesis of the receptor activator of nuclear factor kappa B (RANK) ligand, thus stimulating osteoclast formation and activation.[20,21] Notably, GH induces osteoprotegerin (OPG), which serves as a decoy ligand for RANK, thus blunting effects of IGF-1 on osteoclasts.[22] The effects of GH and IGF-1 on bone are outlined in Fig. 2. In summary, activation of the GH—IGF-1 axis results in an increased bone formation, which is coupled with increased bone resorption and consequently, an increase in bone turnover.[23]

3. GROWTH HORMONE DEFICIENCY IN HUMANS

In addition to the well-established role of GH in promoting linear growth, both GH and IGF-1 influence skeletal turnover, bone size, and bone mineral density (BMD) during childhood and adolescence.[24] In addition to short stature, GH deficient children have smaller bone size and decreased BMD.[25] Whether adults with childhood-onset GH deficiency have lower BMD, is somewhat controversial. A study of 12 patients with isolated childhood-onset GH deficiency, aged 16.3–29.8 years, who had reached skeletal maturation and had discontinued GH replacement at 43.9 months before study entry (range: 9–129 months) found significantly decreased BMD in the lumbar spine, in comparison to an age and gender matched control population (Z score: -1.47, $P < 0.01$), measured by quantitative computed tomography (QCT).[26] Use of this technique yields a true volumetric BMD and avoids introducing a positive bias in the measurement of BMD in patients with small bone size. In contrast, a larger study of adults with childhood-onset GH deficiency, including 18 adults with isolated GH deficiency and 48 adults with multiple pituitary deficiencies, who were almost entirely GH naive, found that BMD Z scores, measured by dual energy X-ray absorptiometry (DXA), were normal in almost all subgroups, after volumetric adjustments were made to account for their small bone size.[27] However, these patients had a markedly increased fracture risk in comparison with a control population, suggesting that BMD may not be accurate in predicting fracture risk in these patients.[27]

Hypopituitary adults with GH deficiency have a decreased bone turnover, as assessed by bone histomorphometry, as well as serum biomarkers of bone formation and resorption.[28,29] In this population, decreased bone turnover results in lower BMD. In a study of 26 hypopituitary patients with an adult-onset GH deficiency of diverse etiologies (hypercortisolism excluded), who had not received GH replacement, median BMD Z scores were significantly lower in comparison with an age- and gender-matched population [Z scores: -1.07 ($P < 0.00005$) in the lumbar spine (measured by QCT), -0.76 ($P = 0.0001$) in the lumbar spine (measured by DXA), and -0.86 ($P = 0.0001$) in the forearm (measured by single-photon absorptiometry)].[30]

As GH deficiency in adults is a heterogeneous syndrome, considerable efforts have been made to identify risk factors for low bone mass (Table 1). Younger hypopituitary adults with GH deficiency are more likely than

Table 1 Risk Factors for Lower BMD in Hypopituitary Adults with GH Deficiency.

Younger age
Increased severity of GH deficiency
• Lower peak GH on stimulation testing
• Lower serum IGF-1 SDS
Central hypogonadism (in the absence of sex steroid replacement)
Central hypoadrenalism

Abbreviations: BMD, bone mineral density; GH, growth hormone; IGF-1, insulin-like growth factor 1; and SDS, standard deviation score.

middle-aged or older patients to have low BMD Z scores.[31] In a study of 125 adults with GH deficiency and hypopituitarism, the proportion of patients with low BMD (Z score <-2.0) was greater in younger patients (lumbar spine: 30% of patients <30 years old vs. 14% of patients >60 years old; femoral neck: 36% of patients <30 years old vs. 0% of patients >60 years old; P for trend < 0.001).[31] In this study, younger patients were more likely than older ones to have childhood-onset GH deficiency or isolated GH deficiency, but were not different with regards to the estimated duration of GH deficiency. These data suggest that low bone mass is relatively infrequent among older hypopituitary patients with GH deficiency. It is possible that some of the younger patients in this study failed to attain optimal peak bone mass, as a consequence of GH deficiency during adolescence.

Other studies have identified the severity of GH deficiency as a risk factor associated with lower BMD in adults.[32,33] In a pharmacoepidemiological study of 1218 hypopituitary adults with GH deficiency of adult onset, who were GH naive at study entry, there was an association between IGF-1 standard deviation scores (SDS) and BMD ($P < 0.001$), which remained robust after adjusting for age, gender, body mass index (BMI), number and type of other pituitary hormone deficiencies.[33] These data are consistent with the hypothesis that GH and IGF-1 have a relevant role in maintaining BMD in adults.[33]

Hypopituitary GH deficient patients may have additional pituitary hormone deficiencies, which might also influence bone health. Indeed, lower BMD T scores in the lumbar spine of unreplaced hypogonadal patients, having GH deficiency (compared with eugonadal GH deficient adults), were reported in another study of 89 hypopituitary adults.[34] In a pharmacoepidemiological study of 1218 adults with GH deficiency, central hypoadrenalism and unreplaced central hypogonadism independently predicted lower BMD in the lumbar spine, consistent with the hypothesis that these hormones and their corresponding replacement therapies may have an important role in bone biology in GH deficient patients.[33]

Despite the importance of BMD in fracture prediction in postmenopausal women and other populations, the value of BMD in predicting fracture risk in patients with GHD is less clear. Several studies have found an increased (two- to threefold) risk of prevalent fracture in hypopituitary adults with GH deficiency, including both morphometric-vertebral and clinical-nonvertebral fractures.[35–38] Smoking was found to increase the risk of prevalent fracture in one study of GH deficient adults.[38] In another study, BMD T scores were not significantly different between patients with prevalent fracture and those without, suggesting that BMD may not accurately predict fracture risk in this population.[35]

The first report on the use of human GH of cadaveric origin, in order to increase height in childhood was published in 1958.[39] At that time, laborious extraction methods were used to isolate GH from human pituitary glands in scarce amounts, which precluded the use of GH as replacement therapy in adults. During the past several decades, the development of recombinant DNA methods that vastly expanded the availability of GH therapy, ultimately led to studies of GH replacement in deficient adults.[2,40,41]

As predicted from *in vitro* data, GH administration to hypopituitary adults increases bone turnover and expands bone remodeling space.[42] Biomarkers of bone formation (osteocalcin, bone-specific alkaline phosphatase, and C-terminal propeptide of type 1 procollagen) and bone resorption (hydroxyproline, deoxypyridinoline, and type 1 collagen N-telopeptide), increase within a period of several months from onset of replacement therapy.[42,43] However, BMD remains unchanged or may even decrease slightly during the first year of GH replacement, followed by an increase in BMD in the lumbar spine and femoral neck.[44,45] The biphasic effect of GH replacement on BMD was confirmed in two meta-analyses, which showed an increase in BMD in response to GH replacement of longer duration (>12–18 months) but no net change in studies of shorter (≤12 months) duration.[44,45] Although another meta-analysis found no increase in BMD after GH replacement, this meta-analysis included a majority (70%) of studies of limited duration (≤6 months), which were combined with studies of longer duration in the analysis.[46] Therefore, the positive effects of GH on BMD, if given for more than 1 year, are supported by published data.

Several predictors of BMD responses to GH replacement have been proposed (Table 2). Gender has been shown to play an important role in

Table 2 Predictors of BMD Response to GH Replacement in Adults.

Patient gender
Duration of therapy
Underlying pituitary disease
Severity of GH deficiency (based on IGF-1 SDS)
Baseline BMD *Z* score

Abbreviations: BMD, bone mineral density; GH, growth hormone; IGF-1, insulin-like growth factor 1; and SDS, standard deviation score.

skeletal response to GH administration in GHD. In aggregate, randomized studies in hypopituitary adults reported that significant increases in BMD (in the lumbar spine and femoral neck) occurred in men, but not in women, receiving GH replacement.[44] These data were corroborated by the findings of prospective studies of long duration (up to 15 years), wherein BMD responses to GH replacement were also greater in men than in women, despite the attainment of higher IGF-1 SDS values in women receiving GH in one study.[47,48] It is unclear if this apparent gender dimorphism is only a consequence of intrinsic differences in skeletal responsiveness to GH replacement or whether it may also be explained on the basis of insufficient GH dosing in women taking oral estrogen in some studies.[49] Women taking oral estrogen require higher GH doses than those on transdermal or no estrogen in order to achieve comparable systemic IGF-1 levels, reflecting the development of GH resistance induced by oral estrogen passing directly through the liver to act on hepatocytes, where most circulating IGF-1 is synthesized.[50,51]

Treatment duration also appears to influence BMD outcomes in GH deficient adults. Long-term prospective studies of GH replacement in adults have shown a sustained, progressive increase in BMD in the lumbar spine (~10% over baseline in men) over a 15 year period.[47,48] In contrast, the effects of GH replacement on BMD in the femoral neck appear to be more modest and less robust.[47,48] In one study of GH replacement in hypopituitary adults, there was a progressive increase in BMD in the femoral neck for 7 years, followed by subsequent decline, suggesting that GH replacement may not prevent age-related bone loss at that site.[48]

The underlying etiology of hypopituitarism may also influence BMD responses to GH replacement, as suggested by the findings of a study of patients with Cushing's disease, prolactinomas, and clinically nonfunctioning pituitary adenomas (NFPA), which reported a delayed increase in BMD in patients with Cushing's disease or prolactinomas, in comparison with the response in those with NFPA.[52] Notably, patients with Cushing's disease or prolactinomas had lower baseline BMD than those with NFPA in this study, which might have also influenced their response to GH replacement.

Other factors that may influence BMD responses to GH replacement include the severity of osteopenia or the extent of GH deficiency at baseline. In a meta-regression of prospective studies of GH replacement in adults, lower baseline *Z* scores (indicating greater severity of osteopenia) or lower IGF-1 SDS (reflecting greater severity of GH deficiency), were associated with a greater BMD increase in the femoral neck.[44] However, neither patient age nor GH dose appear to be independently associated with BMD responses in hypopituitary adults.[44,49,53] Based on a wealth of data demonstrating a relevant role of GH in the skeleton of deficient adults, BMD monitoring by DXA is advisable in adult patients with GH deficiency.[54] The role of bone microarchitecture as a predictor of fracture risk in this population remains to be established.

The potential role of GH replacement influencing peak bone acquisition has also received considerable attention. Patients with idiopathic childhood-onset GH deficiency may often recover their ability to secrete GH as they reach adulthood. In contrast, patients with genetic or structural causes of hypopituitarism are more likely to have persistent GH deficiency past adolescence. Adults with persistent GH deficiency of childhood onset may fail to achieve optimal peak bone mass or experience bone loss if GH replacement does not continue after skeletal maturation occurs.[25] A longer gap between pediatric and adult GH replacement has been associated with lower BMD in adults with childhood-onset GH deficiency persisting in adulthood.[55] Several studies have examined the effects of GH replacement in young adults in transition from adolescence to adulthood over 24 months (Table 3).[56–59] The majority of these studies have found that GH replacement increases BMD in the lumbar spine of patients in transition.[56,58,59] As a corollary, it has been proposed that the gap between pediatric and adult GH replacement be minimized (<24 months) in order to avoid the potentially deleterious consequences arising from interruptions of GH replacement on bone mass in eligible patients, who are found to have persistent GH deficiency on retesting after they reach skeletal maturation.[60] Notably, patients with structural causes of hypopituitarism (such as craniopharyngiomas or pituitary adenomas) and multiple additional pituitary hormone deficiencies, or those with genetic causes of hypopituitarism are likely to have persistent

Table 3 Randomized Controlled Trials of GH Replacement on BMD in the Lumbar Spine in Patients with Persistent GH Deficiency in Transition to Adulthood.

Studies (References)	Duration (in Months)	Effects on BMD (Change Over Baseline)	*P* Value
[59]	24	3.9% (BMD increase)	0.001
[58]	24	2.0–3.0% (BMD increase)	0.027
[57]	24	0.47 (*Z* score increase)	0.086
[56]	24	3.5% (BMD increase)	<0.001

Abbreviations: BMD, bone mineral density; and GH, growth hormone.

GH deficiency into adulthood, and can continue receiving GH replacement during adult life without undergoing repeat GH stimulation testing.[54]

The effects of GH replacement on fracture risk have not been clearly elucidated, owing to a lack of published clinical trials with fracture endpoints. Some retrospective studies suggested that GH replaced adults had a decreased risk of fracture, compared with those who were unreplaced.[35,38] In an observational study from Sweden, including 832 GH deficient adults compared with 2581 matched controls from the general population, the incident fracture risk ratio was significantly lower than 1.0 in adult-onset GH deficient men, suggestive of a protective effect of GH replacement on fracture risk in this subgroup.[61] In adult-onset GH deficient women and childhood-onset GH deficient men, the incident fracture risk ratio was not significantly different from 1.0, suggesting that GH replacement may have mitigated the excess fracture risk in these subgroups.[61] In contrast, childhood-onset GH deficient women had a significantly elevated incident fracture risk ratio [(2.29, 95% confidence intervals: (1.23, 4.28)] despite the institution of GH replacement, possibly reflecting insufficient GH dosing in this subgroup.[61] A recent, large cohort study of 8374 GH replaced hypopituitary adults and 1267 unreplaced GH deficient adults reported a significantly lower fracture incidence rate in the GH replaced (1.19% per year) than the unreplaced (1.91% per year) group, consistent with the hypothesis that GH replacement may mitigate excess fracture risk in deficient adults.[62] However, in the same study there was no protective effect of GH replacement demonstrated in the patient subgroup with osteoporosis at baseline. Randomized clinical trials are needed to resolve this issue definitively, but are unlikely to be conducted, as GH replacement in adults is an approved, widely available therapy in many countries in Europe and the US.

4. GROWTH HORMONE EXCESS AND BONE

GH excess is caused by a pituitary tumor secreting GH autonomously, in >90% cases, and leads to gigantism before epiphyseal fusion and acromegaly during adult life.[63] GH excess leads to increased bone turnover, as evidenced by markers of bone turnover and histomorphometric analysis.[64,65] Bone biomarkers return to normal with biochemical control of GH excess.[66,67] Hypercalciuria is common in active acromegaly and may result from increased calcitriol synthesis, as a consequence of PTH-independent effects of GH and IGF-1 on the kidneys.[68,69] Decreased BMD has been reported in the lumbar spine of patients with acromegaly.[68] Whether this finding is associated with GH excess *per se* or hypogonadism remains controversial. Indeed, other studies have found that decreased BMD in the lumbar spine occurred only in a minority of acromegalic patients and correlated negatively with the presence or duration of hypogonadism.[65,70] Recent data have suggested that trabecular bone microarchitecture may be impaired among eugonadal patients with acromegaly.[71] These findings may explain the observations of an increased prevalence of morphometric-vertebral deformities in patients with acromegaly, which does not appear to be explained by differences in BMD.[72] As discussed in the section on GH deficiency, BMD alone may have limitations in predicting fracture risk in this population. A recent study reported an increased incidence of vertebral fracture in acromegalic patients, particularly those with active disease or hypogonadism.[73] In yet another study, a higher prevalence of vertebral fractures was reported among acromegalic patients with the d3 GH receptor variant, consistent with a relevant role of GH in the pathogenesis of vertebral abnormalities in this population.[74] Acromegalic patients should be evaluated for abnormalities of calcium and vitamin D homeostasis, assessed for hypogonadism, and should also be considered for evaluation of BMD and vertebral structure, particularly if symptomatic (reporting back pain or spine deformity) or hypogonadal.[75,76]

5. GROWTH HORMONE IN NON-DEFICIENT STATES

In addition to GH deficiency, there are several conditions wherein GH therapy may increase growth velocity during childhood leading to greater adult height, including chronic renal failure, Turner syndrome, Prader–Willi syndrome, postnatal growth delay in patients with intrauterine growth retardation who do not demonstrate catch-up growth, idiopathic short stature, short stature homeobox-containing (*SHOX*) gene mutations, and Noonan syndrome.[77–88] Use of GH replacement in these conditions is included among the Food and Drug Administration (FDA)-approved indications for GH replacement during childhood and adolescence (Table 4).

The effects of GH therapy on statural growth, have also been investigated in several other conditions characterized by GH insensitivity. Studies have reported an increase in growth velocity and adult height of children with juvenile rheumatoid arthritis, taking glucocorticoids in pharmacological doses, who also received GH therapy.[89,90] Preliminary studies have reported

Table 4 FDA-Approved Indications for GH Therapy.

Children and Adolescents	Adults
GH deficiency	GH deficiency
Chronic renal failure	HIV associated cachexia
Turner syndrome	Short bowel syndrome
Prader–Willi syndrome	
Intrauterine growth delay without catch-up growth postnatally	
Idiopathic short stature	
SHOX mutation	
Noonan syndrome	

Abbreviations: FDA, Food and Drug Administration; GH, growth hormone; HIV, human immunodeficiency virus; and *SHOX*, short stature homeobox-containing gene.

an increase in growth velocity in response to GH therapy in children with thalassemia or Hurler syndrome, who underwent hematopoietic cell transplantation.[91,92]

Anorexia nervosa is another condition where GH and IGF-1 dysregulation appear to have a pathogenetic role. Anorexia nervosa is characterized by severely low body weight, impaired body image, and intense fear of gaining weight.[93] Anorexia is associated with low BMD, impaired bone microarchitecture, and increased risk of fracture in adolescents and adults.[94–98] In addition to a host of other metabolic and neuroendocrine abnormalities, anorexia nervosa is associated with substantial decreases in serum IGF-1 levels, which correlate with low bone formation biomarkers and decreased BMD despite elevated GH levels, indicative of GH resistance.[99] These observations were further corroborated by the findings of a trial of supraphysiologic-GH therapy in adult women with anorexia nervosa, wherein neither serum IGF-1 levels nor bone biomarkers increased, which was consistent with the presence of GH resistance in this population despite changes in body composition consistent with GH administration.[100] As anticipated from these observations, IGF-1 administration in patients with anorexia nervosa led to increases in biomarkers of bone formation and increased BMD (when given together with oral estrogen–progestin, compared to placebo treated patients).[101,102] Further studies are needed to fully establish the potential therapeutic role of IGF-1 in this population.

GH administration has also been studied in patients with postmenopausal and age-related osteoporosis. Serum IGF-1 levels decline with advancing age as a result of decreased GH secretion.[103] Endogenous GH secretion was reported to correlate positively with BMD values in older women.[104] In older adults with osteoporosis, the effects of GH administration on BMD have been mixed.[105–107] In addition, a meta-analysis of published trials found no benefit of GH administration (with regards to BMD) in older adults with osteoporosis.[108] Thus, available data do not support a therapeutic role for GH in patients with age-related or postmenopausal osteoporosis, but do not exclude the possibility that GH therapy might be of benefit in a subgroup of osteoporotic patients with lower IGF-1 levels.

6. SUMMARY—FUTURE DIRECTIONS

The availability of sensitive assays for GH and IGF-1, the development of non-invasive imaging techniques measuring bone mass and structure, and the application of recombinant methods in the synthesis of GH and IGF-1, leading to the production of these proteins in practically unlimited amounts, have spurred extensive studies on the physiologic role and effects of the GH and IGF-1 axis on the skeleton in diverse conditions.

Available data support a relevant role for GH and IGF-1 as significant factors determining bone mass and structure in conditions characterized by GH deficiency, excess, or insensitivity. Additional studies will be needed to fully elucidate the effects of GH replacement on BMD in women, examine the effects of GH replacement on fracture risk, and characterize the consequences of treatment for acromegaly on the skeleton.

DISCLOSURES

NAT has received research support from Ipsen, Novo Nordisk, and Pfizer. AK has received grant support from Ipsen, Novartis, and Rhythm Pharmaceuticals.

REFERENCES

1. Harris WH, Heaney RP. Effect of growth hormone on skeletal mass in adult dogs. *Nature.* 1969;223(5204):403–404.
2. Goeddel DV, Heyneker HL, Hozumi T, et al. Direct expression in *Escherichia coli* of a DNA sequence coding for human growth hormone. *Nature.* 1979;281(5732):544–548.
3. Daughaday WH. Endocrinology—the way we were: a personal history of somatomedin. *Growth Horm IGF Res.* 2006;16(Suppl A):S3–S5.
4. Salmon Jr WD, Daughaday WH. A hormonally controlled serum factor which stimulates sulfate incorporation by cartilage *in vitro*. *J Lab Clin Med.* 1957;49(6):825–836.

5. Lupu F, Terwilliger JD, Lee K, Segre GV, Efstratiadis A. Roles of growth hormone and insulin-like growth factor 1 in mouse postnatal growth. *Dev Biol.* 2001;229(1):141–162.
6. Ohlsson C, Nilsson A, Isaksson OG, Lindahl A. Effect of growth hormone and insulin-like growth factor-I on DNA synthesis and matrix production in rat epiphyseal chondrocytes in monolayer culture. *J Endocrinol.* 1992;133(2):291–300.
7. Hutchison MR, Bassett MH, White PC. Insulin-like growth factor-I and fibroblast growth factor, but not growth hormone, affect growth plate chondrocyte proliferation. *Endocrinology.* 2007;148(7):3122–3130.
8. Sims NA, Clement-Lacroix P, Da Ponte F, et al. Bone homeostasis in growth hormone receptor-null mice is restored by IGF-I but independent of Stat5. *J Clin Invest.* 2000;106(9): 1095–1103.
9. Zhang M, Xuan S, Bouxsein ML, et al. Osteoblast-specific knockout of the insulin-like growth factor (IGF) receptor gene reveals an essential role of IGF signaling in bone matrix mineralization. *J Biol Chem.* 2002;277(46):44005–44012.
10. Zhao G, Monier-Faugere MC, Langub MC, et al. Targeted overexpression of insulin-like growth factor I to osteoblasts of transgenic mice: increased trabecular bone volume without increased osteoblast proliferation. *Endocrinology.* 2000;141(7):2674–2682.
11. Kassem M, Blum W, Ristelli J, Mosekilde L, Eriksen EF. Growth hormone stimulates proliferation and differentiation of normal human osteoblast-like cells *in vitro*. *Calcif Tissue Int.* 1993;52(3):222–226.
12. Lee JH, Hwang KJ, Kim MY, et al. Human parathyroid hormone increases the mRNA expression of the IGF system and hematopoietic growth factors in osteoblasts, but does not influence expression in mesenchymal stem cells. *J Pediatr Hematol Oncol.* 2012;34(7): 491–496.
13. Watson P, Lazowski D, Han V, Fraher L, Steer B, Hodsman A. Parathyroid hormone restores bone mass and enhances osteoblast insulin-like growth factor I gene expression in ovariectomized rats. *Bone.* 1995;16(3):357–365.
14. Bikle DD, Sakata T, Leary C, et al. Insulin-like growth factor I is required for the anabolic actions of parathyroid hormone on mouse bone. *J Bone Miner Res.* 2002;17(9): 1570–1578.
15. Ernst M, Rodan GA. Estradiol regulation of insulin-like growth factor-I expression in osteoblastic cells: evidence for transcriptional control. *Mol Endocrinol.* 1991;5(8): 1081–1089.
16. Delany AM, Durant D, Canalis E. Glucocorticoid suppression of IGF I transcription in osteoblasts. *Mol Endocrinol.* 2001;15(10):1781–1789.
17. Lakatos P, Caplice MD, Khanna V, Stern PH. Thyroid hormones increase insulin-like growth factor I content in the medium of rat bone tissue. *J Bone Miner Res.* 1993;8(12): 1475–1481.
18. Conover CA. *In vitro* studies of insulin-like growth factor I and bone. *Growth Horm IGF Res.* 2000;10(Suppl B):S107–S110.
19. Clemmons DR, Busby W, Clarke JB, Parker A, Duan C, Nam TJ. Modifications of insulin-like growth factor binding proteins and their role in controlling IGF actions. *Endocr J.* 1998;45(Suppl):S1–S8.
20. Mochizuki H, Hakeda Y, Wakatsuki N, et al. Insulin-like growth factor-I supports formation and activation of osteoclasts. *Endocrinology.* 1992;131(3):1075–1080.
21. Wang Y, Nishida S, Elalieh HZ, Long RK, Halloran BP, Bikle DD. Role of IGF-I signaling in regulating osteoclastogenesis. *J Bone Miner Res.* 2006;21(9):1350–1358.
22. Mrak E, Villa I, Lanzi R, Losa M, Guidobono F, Rubinacci A. Growth hormone stimulates osteoprotegerin expression and secretion in human osteoblast-like cells. *J Endocrinol.* 2007;192(3):639–645.
23. Crane JL, Cao X. Function of matrix IGF-1 in coupling bone resorption and formation. *J Mol Med (Berlin).* 2014;92(2):107–115.

24. Tritos NA, Biller BM. Growth hormone and bone. *Curr Opin Endocrinol Diab Obes.* 2009;16(6):415–422.
25. Baroncelli GI, Bertelloni S, Sodini F, Saggese G. Acquisition of bone mass in normal individuals and in patients with growth hormone deficiency. *J Pediatr Endocrinol Metab.* 2003;16(Suppl 2):327–335.
26. O'Halloran DJ, Tsatsoulis A, Whitehouse RW, Holmes SJ, Adams JE, Shalet SM. Increased bone density after recombinant human growth hormone (GH) therapy in adults with isolated GH deficiency. *J Clin Endocrinol Metab.* 1993;76(5):1344–1348.
27. Bouillon R, Koledova E, Bezlepkina O, et al. Bone status and fracture prevalence in Russian adults with childhood-onset growth hormone deficiency. *J Clin Endocrinol Metab.* 2004;89(10):4993–4998.
28. Bravenboer N, Holzmann P, de Boer H, Blok GJ, Lips P. Histomorphometric analysis of bone mass and bone metabolism in growth hormone deficient adult men. *Bone.* 1996;18(6): 551–557.
29. Sartorio A, Conti A, Monzani M, Morabito F, Faglia G. Growth hormone treatment in adults with GH deficiency: effects on new biochemical markers of bone and collagen turnover. *J Endocrinol Invest.* 1993;16(11):893–898.
30. Holmes SJ, Economou G, Whitehouse RW, Adams JE, Shalet SM. Reduced bone mineral density in patients with adult onset growth hormone deficiency. *J Clin Endocrinol Metab.* 1994;78(3):669–674.
31. Murray RD, Columb B, Adams JE, Shalet SM. Low bone mass is an infrequent feature of the adult growth hormone deficiency syndrome in middle-age adults and the elderly. *J Clin Endocrinol Metab.* 2004;89(3):1124–1130.
32. Colao A, Di Somma C, Pivonello R, et al. Bone loss is correlated to the severity of growth hormone deficiency in adult patients with hypopituitarism. *J Clin Endocrinol Metab.* 1999;84(6):1919–1924.
33. Tritos NA, Greenspan SL, King D, et al. Unreplaced sex steroid deficiency, corticotropin deficiency, and lower IGF-I are associated with lower bone mineral density in adults with growth hormone deficiency: a KIMS database analysis. *J Clin Endocrinol Metab.* 2011;96(5):1516–1523.
34. Mazziotti G, Bianchi A, Cimino V, et al. Effect of gonadal status on bone mineral density and radiological spinal deformities in adult patients with growth hormone deficiency. *Pituitary.* 2008;11(1):55–61.
35. Mazziotti G, Bianchi A, Bonadonna S, et al. Increased prevalence of radiological spinal deformities in adult patients with GH deficiency: influence of GH replacement therapy. *J Bone Miner Res.* 2006;21(4):520–528.
36. Rosen T, Wilhelmsen L, Landin-Wilhelmsen K, Lappas G, Bengtsson BA. Increased fracture frequency in adult patients with hypopituitarism and GH deficiency. *Eur J Endocrinol.* 1997;137(3):240–245.
37. Vestergaard P, Jorgensen JO, Hagen C, et al. Fracture risk is increased in patients with GH deficiency or untreated prolactinomas—a case-control study. *Clin Endocrinol (Oxford).* 2002;56(2):159–167.
38. Wuster C, Abs R, Bengtsson BA, et al. The influence of growth hormone deficiency, growth hormone replacement therapy, and other aspects of hypopituitarism on fracture rate and bone mineral density. *J Bone Miner Res.* 2001;16(2):398–405.
39. Raben MS. Treatment of a pituitary dwarf with human growth hormone. *J Clin Endocrinol Metab.* 1958;18(8):901–903.
40. Salomon F, Cuneo RC, Hesp R, Sonksen PH. The effects of treatment with recombinant human growth hormone on body composition and metabolism in adults with growth hormone deficiency. *N Engl J Med.* 1989;321(26):1797–1803.
41. Jorgensen JO, Pedersen SA, Thuesen L, et al. Beneficial effects of growth hormone treatment in GH-deficient adults. *Lancet.* 1989;1(8649):1221–1225.

42. Hansen TB, Brixen K, Vahl N, et al. Effects of 12 months of growth hormone (GH) treatment on calciotropic hormones, calcium homeostasis, and bone metabolism in adults with acquired GH deficiency: a double blind, randomized, placebo-controlled study. *J Clin Endocrinol Metab.* 1996;81(9):3352–3359.
43. Baum HB, Biller BM, Finkelstein JS, et al. Effects of physiologic growth hormone therapy on bone density and body composition in patients with adult-onset growth hormone deficiency. A randomized, placebo-controlled trial. *Ann Intern Med.* 1996;125(11): 883–890.
44. Barake M, Klibanski A, Tritos NA. Effects of recombinant human growth hormone therapy on bone mineral density in adults with growth hormone deficiency: a meta-analysis. *J Clin Endocrinol Metab.* 2014;99(3):852–860.
45. Davidson P, Milne R, Chase D, Cooper C. Growth hormone replacement in adults and bone mineral density: a systematic review and meta-analysis. *Clin Endocrinol (Oxford).* 2004;60(1):92–98.
46. Hazem A, Elamin MB, Bancos I, et al. Body composition and quality of life in adults treated with GH therapy: a systematic review and meta-analysis. *Eur J Endocrinol.* 2012;166(1): 13–20.
47. Appelman-Dijkstra NM, Claessen KM, Hamdy NA, Pereira AM, Biermasz NR. Effects of up to 15 years of recombinant human GH (rhGH) replacement on bone metabolism in adults with growth hormone deficiency (GHD): the Leiden Cohort Study. *Clin Endocrinol (Oxford).* 2014;81(5):727–735.
48. Elbornsson M, Gotherstrom G, Bosaeus I, Bengtsson BA, Johannsson G, Svensson J. Fifteen years of GH replacement increases bone mineral density in hypopituitary patients with adult-onset GH deficiency. *Eur J Endocrinol.* 2012;166(5):787–795.
49. Tritos NA, Hamrahian AH, King D, et al. Predictors of the effects of 4 years of growth hormone replacement on bone mineral density in patients with adult-onset growth hormone deficiency—a KIMS database analysis. *Clin Endocrinol (Oxford).* 2013;79(2): 178–184.
50. Cook DM. Growth hormone and estrogen: a clinician's approach. *J Pediatr Endocrinol Metab.* 2004;17(Suppl 4):1273–1276.
51. Cook DM, Ludlam WH, Cook MB. Route of estrogen administration helps to determine growth hormone (GH) replacement dose in GH-deficient adults. *J Clin Endocrinol Metab.* 1999;84(11):3956–3960.
52. Colson A, Brooke AM, Walker D, et al. Growth hormone deficiency and replacement in patients with treated Cushing's Disease, prolactinomas and non-functioning pituitary adenomas: effects on body composition, glucose metabolism, lipid status and bone mineral density. *Horm Res.* 2006;66(6):257–267.
53. Elbornsson M, Gotherstrom G, Franco C, Bengtsson BA, Johannsson G, Svensson J. Effects of 3-year GH replacement therapy on bone mineral density in younger and elderly adults with adult-onset GH deficiency. *Eur J Endocrinol.* 2012;166(2): 181–189.
54. Molitch ME, Clemmons DR, Malozowski S, Merriam GR, Vance ML, Endocrine S. Evaluation and treatment of adult growth hormone deficiency: an Endocrine Society clinical practice guideline. *J Clin Endocrinol Metab.* 2011;96(6):1587–1609.
55. Tritos NA, Hamrahian AH, King D, et al. A longer interval without GH replacement and female gender are associated with lower bone mineral density in adults with childhood-onset GH deficiency: a KIMS database analysis. *Eur J Endocrinol.* 2012;167 (3):343–351.
56. Conway G, Szarras-Czapnik M, Racz K, et al. Treatment for 24 months with recombinant human growth hormone has a beneficial effect on bone mineral density in young adults with childhood-onset growth hormone deficiency. *Eur J Endocrinol.* 2009;160 (6):899–907.

57. Mauras N, Pescovitz OH, Allada V, Messig M, Wajnrajch MP, Lippe B. Limited efficacy of growth hormone (GH) during transition of GH-deficient patients from adolescence to adulthood: a phase III multicenter, double-blind, randomized two-year trial. *J Clin Endocrinol Metab.* 2005;90(7):3946–3955.
58. Shalet SM, Shavrikova E, Cromer M, et al. Effect of growth hormone (GH) treatment on bone in postpubertal GH-deficient patients: a 2-year randomized, controlled, dose-ranging study. *J Clin Endocrinol Metab.* 2003;88(9):4124–4129.
59. Underwood LE, Attie KM, Baptista J. Growth hormone (GH) dose-response in young adults with childhood-onset GH deficiency: a two-year, multicenter, multiple-dose, placebo-controlled study. *J Clin Endocrinol Metab.* 2003;88(11):5273–5280.
60. Nguyen VT, Misra M. Transitioning of children with GH deficiency to adult dosing: changes in body composition. *Pituitary.* 2009;12(2):125–135.
61. Holmer H, Svensson J, Rylander L, et al. Fracture incidence in GH-deficient patients on complete hormone replacement including GH. *J Bone Miner Res.* 2007;22(12): 1842–1850.
62. Mo D, Fleseriu M, Qi R, et al. Fracture risk in adult patients treated with growth hormone replacement therapy for growth hormone deficiency: a prospective observational cohort study. *Lancet Diab Endocrinol.* 2015;3(5):331–338.
63. Katznelson L, Laws Jr ER, Melmed S, et al. Acromegaly: an endocrine society clinical practice guideline. *J Clin Endocrinol Metab.* 2014;99(11):3933–3951.
64. Ueland T, Fougner SL, Godang K, Schreiner T, Bollerslev J. Serum GH and IGF-I are significant determinants of bone turnover but not bone mineral density in active acromegaly: a prospective study of more than 70 consecutive patients. *Eur J Endocrinol.* 2006;155(5):709–715.
65. Bolanowski M, Daroszewski J, Medras M, Zadrozna-Sliwka B. Bone mineral density and turnover in patients with acromegaly in relation to sex, disease activity, and gonadal function. *J Bone Miner Metab.* 2006;24(1):72–78.
66. Fairfield WP, Sesmilo G, Katznelson L, et al. Effects of a growth hormone receptor antagonist on bone markers in acromegaly. *Clin Endocrinol (Oxford).* 2002;57(3):385–390.
67. Parkinson C, Kassem M, Heickendorff L, Flyvbjerg A, Trainer PJ. Pegvisomant-induced serum insulin-like growth factor-I normalization in patients with acromegaly returns elevated markers of bone turnover to normal. *J Clin Endocrinol Metab.* 2003;88 (12):5650–5655.
68. Ezzat S, Melmed S, Endres D, Eyre DR, Singer FR. Biochemical assessment of bone formation and resorption in acromegaly. *J Clin Endocrinol Metab.* 1993;76(6):1452–1457.
69. Kamenicky P, Mazziotti G, Lombes M, Giustina A, Chanson P. Growth hormone, insulin-like growth factor-1, and the kidney: pathophysiological and clinical implications. *Endocr Rev.* 2014;35(2):234–281.
70. Kayath MJ, Vieira JG. Osteopenia occurs in a minority of patients with acromegaly and is predominant in the spine. *Osteoporos Int.* 1997;7(3):226–230.
71. Madeira M, Neto LV, de Paula Paranhos Neto F, et al. Acromegaly has a negative influence on trabecular bone, but not on cortical bone, as assessed by high-resolution peripheral quantitative computed tomography. *J Clin Endocrinol Metab.* 2013;98(4): 1734–1741.
72. Mazziotti G, Bianchi A, Bonadonna S, et al. Prevalence of vertebral fractures in men with acromegaly. *J Clin Endocrinol Metab.* 2008;93(12):4649–4655.
73. Mazziotti G, Bianchi A, Porcelli T, et al. Vertebral fractures in patients with acromegaly: a 3-year prospective study. *J Clin Endocrinol Metab.* 2013;98(8):3402–3410.
74. Mormando M, Nasto LA, Bianchi A, et al. GH receptor isoforms and skeletal fragility in acromegaly. *Eur J Endocrinol.* 2014;171(2):237–245.
75. Giustina A, Bronstein MD, Casanueva FF, et al. Current management practices for acromegaly: an international survey. *Pituitary.* 2011;14(2):125–133.

76. Melmed S, Casanueva FF, Klibanski A, et al. A consensus on the diagnosis and treatment of acromegaly complications. *Pituitary.* 2013;16(3):294–302.
77. Haffner D, Schaefer F, Nissel R, Wuhl E, Tonshoff B, Mehls O. Effect of growth hormone treatment on the adult height of children with chronic renal failure. German Study Group for Growth Hormone Treatment in Chronic Renal Failure. *N Engl J Med.* 2000;343(13):923–930.
78. Mehls O, Lindberg A, Nissel R, Haffner D, Hokken-Koelega A, Ranke MB. Predicting the response to growth hormone treatment in short children with chronic kidney disease. *J Clin Endocrinol Metab.* 2010;95(2):686–692.
79. Mehls O, Wuhl E, Tonshoff B, Schaefer F, Nissel R, Haffner D. Growth hormone treatment in short children with chronic kidney disease. *Acta Paediatr.* 2008;97(9): 1159–1164.
80. Nissel R, Lindberg A, Mehls O, Haffner D. Pfizer International Growth Database International B.Factors predicting the near-final height in growth hormone-treated children and adolescents with chronic kidney disease. *J Clin Endocrinol Metab.* 2008;93(4): 1359–1365.
81. Aycan Z, Cetinkaya E, Darendeliler F, et al. The effect of growth hormone treatment on bone mineral density in prepubertal girls with Turner syndrome: a multicentre prospective clinical trial. *Clin Endocrinol (Oxford).* 2008;68(5):769–772.
82. Davenport ML, Crowe BJ, Travers SH, et al. Growth hormone treatment of early growth failure in toddlers with Turner syndrome: a randomized, controlled, multicenter trial. *J Clin Endocrinol Metab.* 2007;92(9):3406–3416.
83. Argente J, Gracia R, Ibanez L, et al. Improvement in growth after two years of growth hormone therapy in very young children born small for gestational age and without spontaneous catch-up growth: results of a multicenter, controlled, randomized, open clinical trial. *J Clin Endocrinol Metab.* 2007;92(8):3095–3101.
84. Willemsen RH, Arends NJ, Bakker-van Waarde WM, et al. Long-term effects of growth hormone (GH) treatment on body composition and bone mineral density in short children born small-for-gestational-age: six-year follow-up of a randomized controlled GH trial. *Clin Endocrinol (Oxford).* 2007;67(4):485–492.
85. Kemp SF, Kuntze J, Attie KM, et al. Efficacy and safety results of long-term growth hormone treatment of idiopathic short stature. *J Clin Endocrinol Metab.* 2005;90(9): 5247–5253.
86. Moore WV, Dana K, Frane J, Lippe B. Growth hormone responsiveness: peak stimulated growth hormone levels and other variables in idiopathic short stature (ISS): data from the National Cooperative Growth Study. *Pediatr Endocrinol Rev.* 2008;6(1):5–8.
87. Blum WF, Cao D, Hesse V, et al. Height gains in response to growth hormone treatment to final height are similar in patients with SHOX deficiency and Turner syndrome. *Horm Res.* 2009;71(3):167–172.
88. Romano AA, Dana K, Bakker B, et al. Growth response, near-adult height, and patterns of growth and puberty in patients with Noonan syndrome treated with growth hormone. *J Clin Endocrinol Metab.* 2009.
89. Simon D, Prieur AM, Quartier P, Charles Ruiz J, Czernichow P. Early recombinant human growth hormone treatment in glucocorticoid-treated children with juvenile idiopathic arthritis: a 3-year randomized study. *J Clin Endocrinol Metab.* 2007;92(7): 2567–2573.
90. Bechtold S, Ripperger P, Dalla Pozza R, et al. Growth hormone increases final height in patients with juvenile idiopathic arthritis: data from a randomized controlled study. *J Clin Endocrinol Metab.* 2007;92(8):3013–3018.
91. Geffner ME, Karlsson H. Use of recombinant human growth hormone in children with thalassemia. *Horm Res.* 2009;71(Suppl 1):46–50.

92. Polgreen LE, Plog M, Schwender JD, et al. Short-term growth hormone treatment in children with Hurler syndrome after hematopoietic cell transplantation. *Bone Marrow Transpl.* 2009.
93. Smink FR, van Hoeken D, Hoek HW. Epidemiology, course, and outcome of eating disorders. *Curr Opin Psychiatry.* 2013;26(6):543–548.
94. Misra M, Aggarwal A, Miller KK, et al. Effects of anorexia nervosa on clinical, hematologic, biochemical, and bone density parameters in community-dwelling adolescent girls. *Pediatrics.* 2004;114(6):1574–1583.
95. Biller BM, Saxe V, Herzog DB, Rosenthal DI, Holzman S, Klibanski A. Mechanisms of osteoporosis in adult and adolescent women with anorexia nervosa. *J Clin Endocrinol Metab.* 1989;68(3):548–554.
96. Faje AT, Karim L, Taylor A, et al. Adolescent girls with anorexia nervosa have impaired cortical and trabecular microarchitecture and lower estimated bone strength at the distal radius. *J Clin Endocrinol Metab.* 2013;98(5):1923–1929.
97. Faje AT, Fazeli PK, Miller KK, et al. Fracture risk and areal bone mineral density in adolescent females with anorexia nervosa. *Int J Eat Disord.* 2014;47(5):458–466.
98. Lawson EA, Miller KK, Bredella MA, et al. Hormone predictors of abnormal bone microarchitecture in women with anorexia nervosa. *Bone.* 2010;46(2):458–463.
99. Misra M, Miller KK, Bjornson J, et al. Alterations in growth hormone secretory dynamics in adolescent girls with anorexia nervosa and effects on bone metabolism. *J Clin Endocrinol Metab.* 2003;88(12):5615–5623.
100. Fazeli PK, Lawson EA, Prabhakaran R, et al. Effects of recombinant human growth hormone in anorexia nervosa: a randomized, placebo-controlled study. *J Clin Endocrinol Metab.* 2010;95(11):4889–4897.
101. Grinspoon S, Thomas L, Miller K, Herzog D, Klibanski A. Effects of recombinant human IGF-I and oral contraceptive administration on bone density in anorexia nervosa. *J Clin Endocrinol Metab.* 2002;87(6):2883–2891.
102. Misra M, McGrane J, Miller KK, et al. Effects of rhIGF-1 administration on surrogate markers of bone turnover in adolescents with anorexia nervosa. *Bone.* 2009;45(3): 493–498.
103. Landin-Wilhelmsen K, Wilhelmsen L, Lappas G, et al. Serum insulin-like growth factor I in a random population sample of men and women: relation to age, sex, smoking habits, coffee consumption and physical activity, blood pressure and concentrations of plasma lipids, fibrinogen, parathyroid hormone and osteocalcin. *Clin Endocrinol (Oxford).* 1994;41(3):351–357.
104. Dennison EM, Hindmarsh PC, Kellingray S, Fall CH, Cooper C. Growth hormone predicts bone density in elderly women. *Bone.* 2003;32(4):434–440.
105. Gillberg P, Mallmin H, Petren-Mallmin M, Ljunghall S, Nilsson AG. Two years of treatment with recombinant human growth hormone increases bone mineral density in men with idiopathic osteoporosis. *J Clin Endocrinol Metab.* 2002;87(11):4900–4906.
106. Landin-Wilhelmsen K, Nilsson A, Bosaeus I, Bengtsson BA. Growth hormone increases bone mineral content in postmenopausal osteoporosis: a randomized placebo-controlled trial. *J Bone Miner Res.* 2003;18(3):393–405.
107. Sugimoto T, Kaji H, Nakaoka D, et al. Effect of low-dose of recombinant human growth hormone on bone metabolism in elderly women with osteoporosis. *Eur J Endocrinol.* 2002;147(3):339–348.
108. Liu H, Bravata DM, Olkin I, et al. Systematic review: the safety and efficacy of growth hormone in the healthy elderly. *Ann Intern Med.* 2007;146(2):104–115.

INDEX

A

E

F

G

H

I

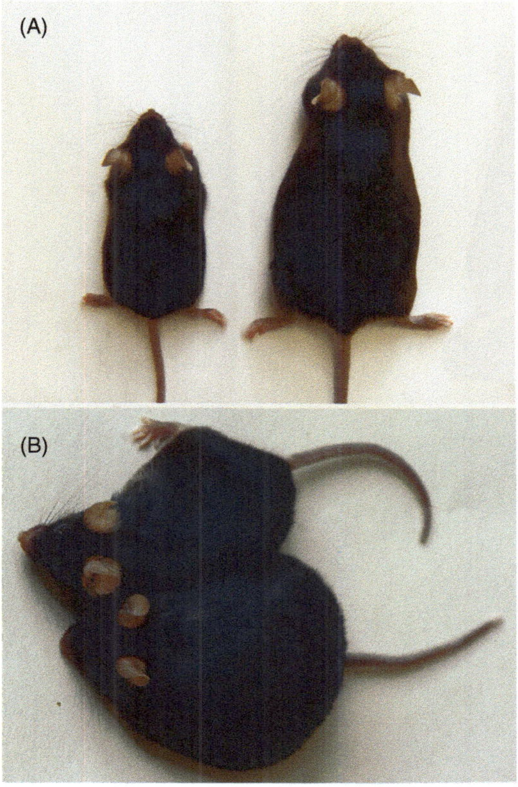

Jonathan A. Young *et al.*, Figure 1 Images of global GHR−/− and fat-specific GHRKO mice. (A) 6-month-old global GHR−/− mouse (left) next to a littermate control (right). (B) 6-month-old FaGHRKO (bottom) and littermate control (top).

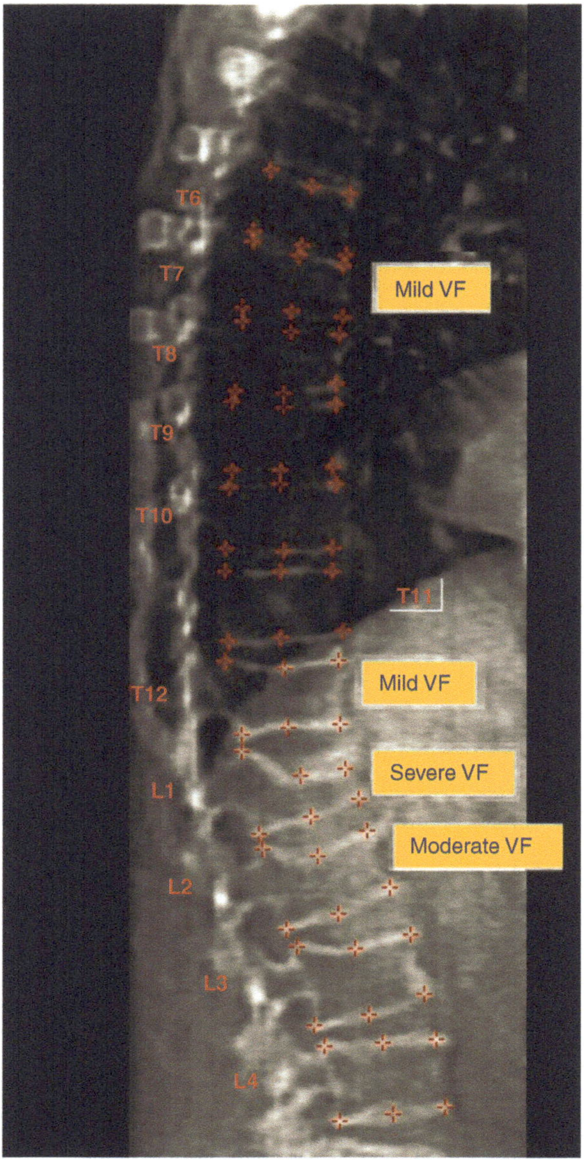

Stefano Frara *et al.*, Figure 1 Lateral DXA image of thoracic and lumbar spine showing four VFs in an acromegaly patient (T7 and T12 mild, L2 moderate, L1 severe; according to Genant classification; see text for details); the fractures were identified by marking the vertebral bodies with six points to define the vertebral shape and heights.

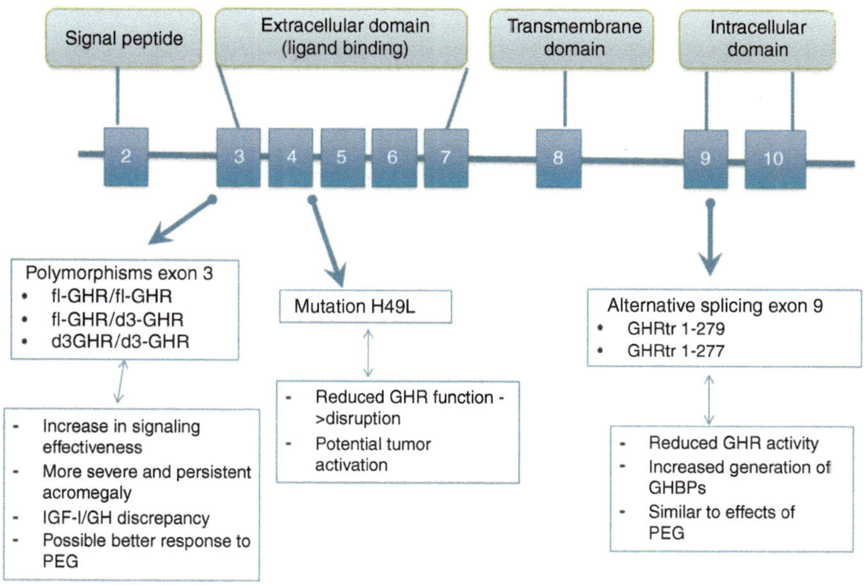

Ana M. Ramos-Leví *et al.*, Figure 1 Schematic representation of the GHR, in the short arm of chromosome (p13.1-p12).[72] This gene comprises 9 exons, which encode the signal peptide, the extracellular domain, and the intracellular domain. Variants of the GHR: deletion of exon 3, H49L mutation, and truncated forms due to alternative splicing of exon 9.

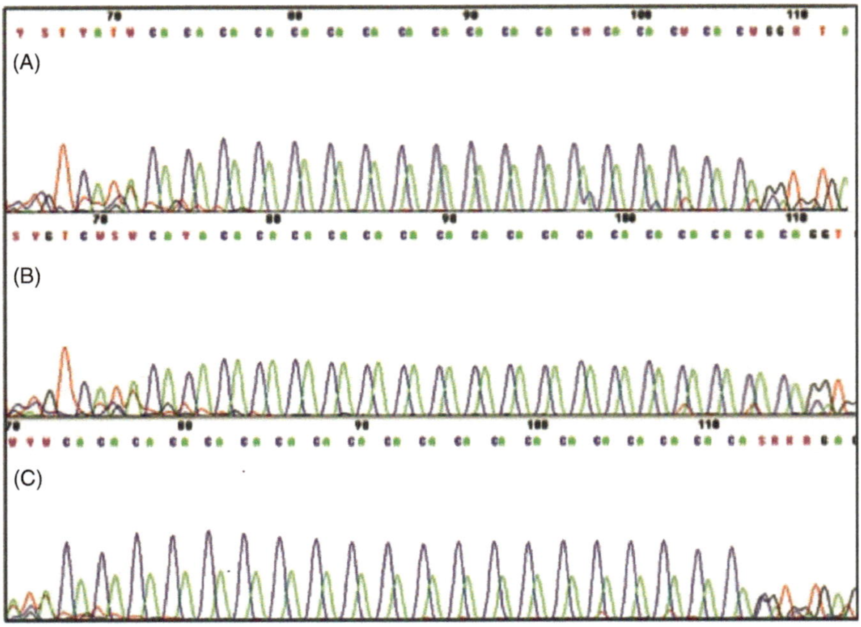

Ana M. Ramos-Leví *et al.*, Figure 2 IGF-I(CA) polymorphism. Examples from patients with (A) 18 CA repeats, (B) 19 CA repeats, and (C) 20 CA repeats.

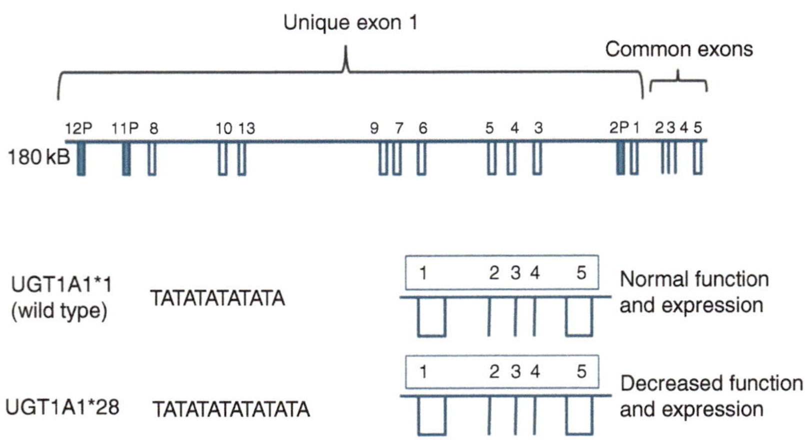

Ana M. Ramos-Leví *et al.*, Figure 3 Schematic representation of the UGT1A1 polymorphism.

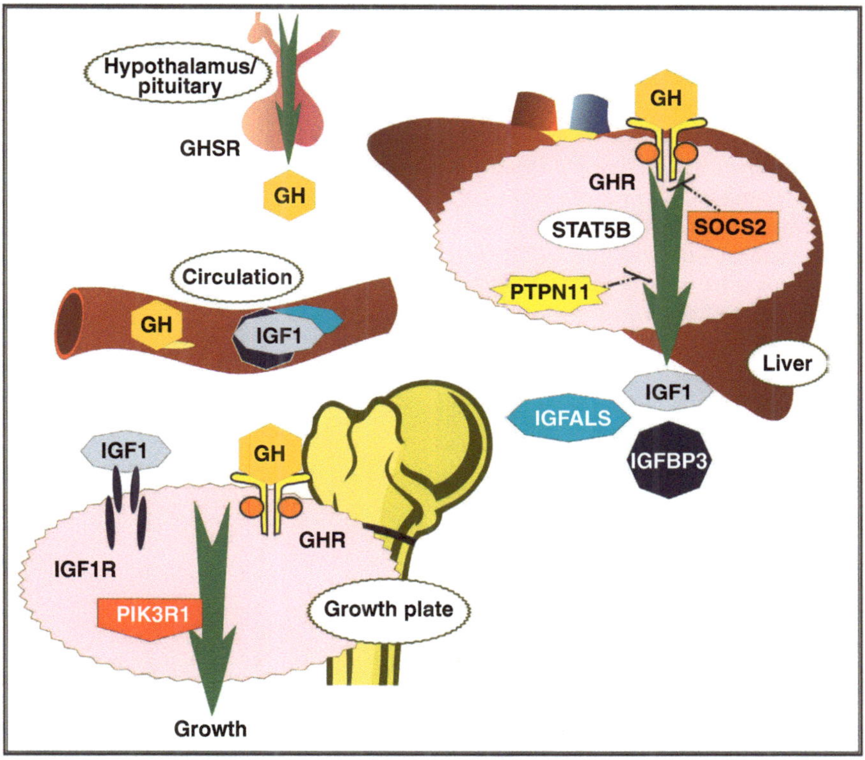

Hugo L. Fideleff *et al.*, Figure 1 GH/IGF1 axis: schematic representation. Pituitary GH secretion, intracellular hepatic mechanisms for IGF1 production, transportation as a ternary complex in circulation and direct action of IGF1 and GH in growth plate are depicted. Abbreviations: GH, growth hormone; GHSR, growth hormone secretagogue receptor; GHR, growth hormone receptor; STAT5B, signal transducer and activator of transcription 5B; SOCS2, suppressor of cytokine signaling 2; PTPN11, tyrosine-protein phosphatase nonreceptor type 11; IGF1, insulin-like growth factor 1; IGFALS, IGF acid labile subunit; IGFBP3, insulin-like growth factor binding protein 3; IGF1R, insulin-like growth factor 1 receptor; PIK3R1, phosphatidylinositol 3-kinase regulatory subunit alpha.

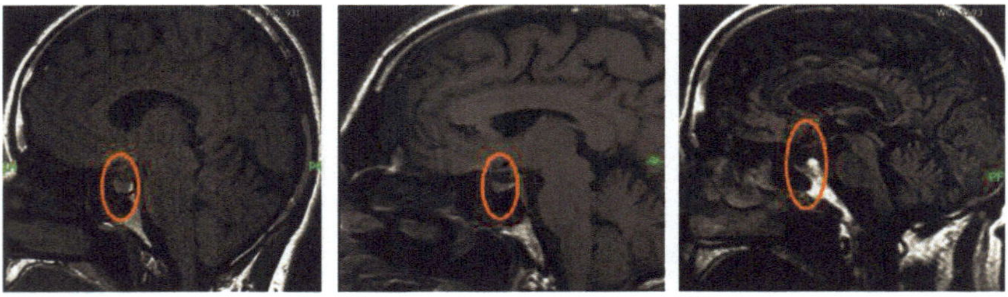

(A) (B) (C)

Züleyha Karaca *et al.*, Figure 1 Three-dimensional MRI of a healthy control (A), 48 year normal control pituitary volume: 717 mm^3; GH-sufficient retired boxer (B), 46 year retired boxer without GH deficiency pituitary volume: 631 mm^3; and GH-deficient retired boxer (C), 47 year retired boxer with GH deficiency pituitary volume: 267 mm^3. GH-deficient retired boxer has significantly smaller pituitary volume than healthy control and GH-sufficient retired boxer.

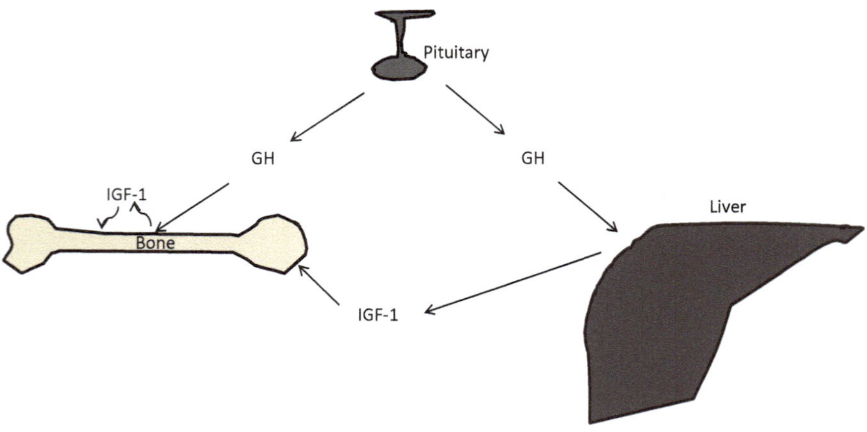

Nicholas A. Tritos and Anne Klibanski, Figure 1 Growth hormone, insulin-like growth factor 1 (IGF-1), and the skeleton. Abbreviations: GH, growth hormone; and IGF-1, insulin-like growth factor 1.

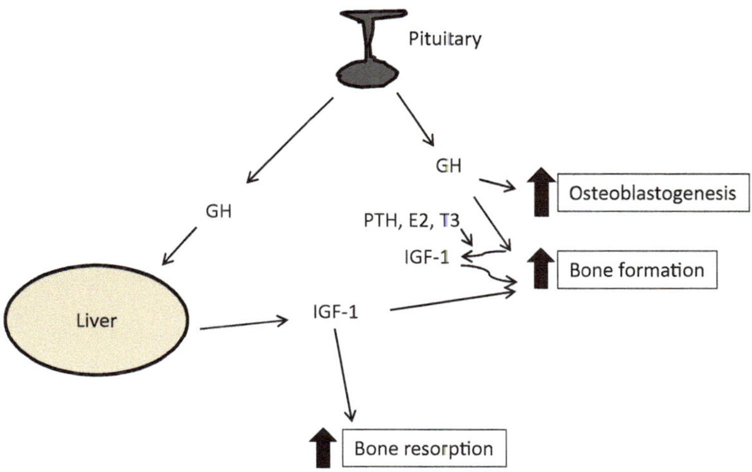

Nicholas A. Tritos and Anne Klibanski, Figure 2 Effects of growth hormone, insulin-like growth factor 1 (IGF-1), and other hormones on skeletal tissue. Abbreviations: E2, estradiol; GH, growth hormone; IGF-1, insulin-like growth factor 1; PTH, parathyroid hormone; and T3, triiodothyronine.

FSC
www.fsc.org
MIX
Papier | Fördert
gute Waldnutzung
FSC® C083411

Kolibri 360 GmbH
Ferdinand-Jühlke-Straße 7
99095 Erfurt, Deutschland
produktsicherheit@kolibri360.de